Dark Age Avengers

By

Marcel Rene Chenard

Edited by

Austin Mardon

Golden Meteorite Press

A Golden Meteorite Press Book.

Printed in Canada by Espresso Books.

Previously published by Golden Meteorite Press under the titles *The Avengers Trilogy* – 2003, *Samson's Return* – 2004, *Beyond Galala Valley* – 2007, *Back To Euta Vicinity* – 2008, *In The Aftermath* – 2009.

Supported by grant from Antarctic Institute of Canada.

Cover art taken from Bayeux Tapestry, 1080's AD.
Cover design Leah Marcus
Published by Golden Meteorite Press.
Post Office Box 1223, Station Main,
Edmonton, Alberta, Canada. T5B 2W4
Telephone: 780-378-0063

ISBN 978-1-897480-06-9

Library and Archives Canada Cataloguing in Publication

Chenard, Marcel Rene, 1967-
Dark age avengers / by Marcel Rene Chenard ; Austin Mardon, editor. -- 2nd ed.

ISBN 978-1-897480-06-9

I. Mardon, Austin A. (Austin Albert) II. Title.

PS8555.H44512D37 2010 C813'.6 C2010-900006-4

Dedicated
To
Micah Johnson

Dark Age Avengers

Book 1 The Avengers:Introduction

In my teens I used to play Dungeons & Dragons with my friend Robert who was also in his teens. He was native of origin and a great fellow to make company with. We created adventures which were both moving and sometimes scary. What follows is the result of our efforts in an improvised dramatization of our charecters' experiences in novel format. Somewhat like Dragon Lance. I am writing this for the entertainment of my interested readers.

The characters used are named after the original characters which we created together. Sassy and Samson are my own design as Suchi and the fighters are Robert's creations. I hope you will enjoy the adventures within, so now read and enjoy.

Marcel Rene Chenard, The Author.

Chapter 1

Samson Leader entered the Boars Head Inn and approached the bar. He was a dwarven creature of average height and his eyes glowed with life, like two orbs of blue topaz. He had been through a lot in his life and had learned to put on a poker face like the rest of the town's gamblers, as it was needed in his line of work as well as in gaming. His expression was one of neutrality as he dragged his feet in exhaustion. He had just returned from a long trek. He didn't call himself Samson Leader for nothing.

He had been born Samson of Clan Trellis but as he was a Dwarven guide he called himself Samson Leader. As much to hide his ancestry as anything else. As he placed five copper pieces down at the bar he asked for a mug of ale.

Isolating himself at a table he took long swigs on his ale to quench his thirst.

Then he placed the empty mug on the table and began to scan the room. Two fighters sat at a table across the way but otherwise the inn was empty. It was still midmorning.

Colina, what Samson considered to be his home town, sat along the westwardly winding river five days' journey from the ocean-front town of Anasthasia. He was glad he had made it here unscathed. He had guided several dignitaries to Colina from the ocean-front community and was glad there had been no incidence. But he recalled now that he would have preferred some action. His group had been fairly large which had promoted the peaceful trek. Hellhouds did not attack large groups and he knew that they would have stayed far in the distance. At any rate the five day trek had been quite boring. Or it would have been if not for the conversation.

Now he sat back and relaxed, resting his short but muscular legs and rubbing his hand through his bristly beard. He recalled the chatter of the previous night and noticed again that the sole woman on the trek was in large demand. She would indeed make a fine mate for the prince.

Leo Anasthasius was in charge of the entire principality as far as the Calico Mountains far to the north. The trek to the mountains had never been recorded as the mountains were considered impossible to traverse. No one had even considered traveling on such a grand adventure. Maybe some day Samson would attempt it.

The prince was due to arrive in Colina the following day. Then he would join with the young lady that Samson had conversed with about local politics. She had a unique knowledge of the land and its people. This would make her the ideal person to sit at the prince's right hand.

Unfailingly his mind turned to the destiny of Clan Trellis. His clan was all but extinct because of the Clan Wars which continued to rage in the Dwarven Territories. He recalled himself as a young man watching his treasured mother fall from the attack of a marauding clan. He would not fight his own people. Some one had put him away from the battle and placed him in sanctuary at a neighboring dwelling of a supporting clan. Later he would be shipped out for his own safety. Samson was the prince of Clan Trellis. Son of the headman of his clan. But he would never reveal his ancestry lest a warring opponent find him and establish his death. Some day, when the time was right, with a woman at his side, he would reform his own clan as the headman of Clan Trellis. But not until all the warring factions had died out. That was his mission in life. To rejuvenate the clan with the woman of his choice. Which, unfortunately, he hadn't found yet. But not many dwarves stayed in this area of the Principality of the Crane. Most tried to struggle through the terrible blood shed in the Dwarven Territories to the north-east. But he would not return. Not yet. Maybe some day, when the war was over, he would return to claim his title. But not until the warring factions died out.

As he stepped up to the bar to replenish his drink a human cleric entered into his presence. He was a good looking chap to Samson's standards, with dirty-blonde hair and with blue eyes that he knew a woman would sink into. He knew that this man was well-known in this district. But he had never had the pleasure of his acquaintance before. He was a loner, Samson was, as was best for his life's sake. The young-looking, clean shaven cleric walked up to the bar and sat down on a seat next to Samson. Turning, he spoke.

"Your name is Samson Leader, is it not?"

"Yes, I am him. And who might you be?"

"I am Suchi Evil Killer, acolyte to Lord Brunweger. I keep hearing good things about you. You're the one who brought the princess to town, I hear."

"It's spreading already? I just got here a few minutes ago." The bartender poured Samson another drink and was passed five more pieces. "I'm not famous just for one job, you know. I'm well-known in these parts."

"That is why I have come. I have need of your assistance."

"I'm listening."

"His Lordship Lord Brunweger wishes me to organize a permanent group of adventurers. I come to you because you are the best. You've even slaughtered many a hellhound in your time. I sometimes wonder why you are still with us."

"I am best at slaughtering those fire-breathing animals. But they are only animals. I have seen many an amateur panic in sight of a fire-breathing creature. I have only once encountered a chloroform-breathing dragon. Luckily, I was never in danger. "

"Join with me and my band of merry men. You are the first but I know who I want."

"I am at your service, my friend."

"His Lordship will sponsor this group and provide lodging as well as any supplies we may need. You are in good hands with Lord Eugene Brunweger. He is now keeping court with the princess until his grandeur the prince arrives in town tomorrow."

"I have been appraised. It seems we have labour ahead of us hunting up these merry men. Shall we be off?"

"Finish your ale, my friend! I will have further word with you. I also have word that we have our first adventure ahead of us. So we must work quickly. His Lordship has heard that an Elven Temple has been captured by an evil magic user. He is just now attempting to find the village's hidden treasures. And he is using, tormenting, the village's headman as a result. We must be prompt. We are to travel to Village Wildlife in the Elven Territories and talk to the Elven maiden named Sasselia Wildlife. She is a powerful magic user and can conquer the human magic user. She thinks. But she needs our support. I am looking forward to meeting the beautiful maiden. Perhaps she will decide to join our group also, I must ask." Samson emptied his second ale.

"Let's be off, then."

Suchi led the way into the wayfare of the town, down several side streets, and entered the local fighters' guild. Entering into the presence of a stout, elderly gentleman behind a desk, he said a few words, and the stout man went through a door.

"The Bloodbath brothers are the best in the fighter guild. They have seen lots of action. I am sure they will be glad to join us."

"Suchi!" someone exclaimed as two men came into their presence.

"Samson, this is Roger and his brother Roderick." The Bloodbath brothers were as much alike as two men could be. Roger was clean shaven and had a balding round of hair on his scalp. His brother was younger-looking and had a bristly beard just beginning on his face. Both brothers had deep brown eyes and an ivory complexion.

"I have come to ask you to join a band of men His Lordship Brunweger wishes to organize. It will be a permanent establishment under him which will destroy the great evil in this land. He'll sponsor us and provide lodging, supplies, anything we may need. We will be a grand group of merry men. And we will be the best, I guarantee you. What do you say?"

"We're in!" both brothers exclaimed at once. Roger continued. "Roderick has just completed his apprenticeship and is now a licensed swordsman. And a unique fighter in his own right."

"Can you join us on a short trek?"

"Where to?"

"The local thieves' guild. Our presence may cause some trouble."

"Understood. We are armed as you can see. I thought you might need us."

"On then."

Suchi led the way with Samson directly behind him as the two fighters followed. Suchi knew they had a long trek ahead of them, the river front district was the industrial centre of Colina, but the streets were full of backstabbing thieves and murderous half-humans, mutated by the lust of inter-species mating. This was a vulgar district and the men knew they must be careful and keep their eyes open at all times.

"Gentlemen, I would like to introduce my dear friend, Tralina."

"Where did you find her, Thonolan? Looks like you're getting soft in your old age. Time to settle down and have a clan of little mutants?"

"Humans and Elves have always been known to be sexually compatible."

"Really? Well, maybe I will show her how to have a good time." Thonolan found himself in a really tight bind. His fellows were making a complete sham of him, and he didn't like it, either. They always found ways to tease or threaten him into doing their bidding. But not this time!

"She's my fiance!"

"Which will make me even happier to take her off your hands."

Thonolan pulled out his dagger in storming rage and pushed Tralina into a corner. The expression on the elven maiden's face revealed her fear, not for herself, but for the man she loved. Thonolan jabbed the air with his dagger and approached his opponents in swift strides. The blade moved closer and closer. When he was within reach of one, the chaotic thief moved aside and grabbed Thonolan's two arms, forcing them behind his back, and putting him in a neck hold. Revealing the tender flesh of his young neck, the opponent held Thonolan in a grasp. Another man pulled out a dagger and approached Thonolan threateningly.

Suchi heard a female scream. The two fighters, the cleric and the dwarf rushed down a flight of stairs to the rescue. Entering into the scene, the fighter drew their swords. Samson, being a dwarf, drew the only weapon he possessed. His father's dagger.

Roger saw the problem at first glance. Thieves were about to slaughter one of their own. He ran toward the thieves and back stabbed the one with the dagger held to slit Thonolan's throat. The blade went through one side and came out his gorged belly. The thief buckled and collapsed. The thief behind Thonolan pushed him aside and attempted to stab Roger with a frontal assault. But this fighter, who had years of experience, had tremendous dexterity. He rushed aside, and as the thief flung at thin air, he slit the man's throat at the jugular. Blood began to pour as the chaotic thief tripped and fell. He gasped for breath as he saw his own blood flow.

While Roger had been attacking the attackers more thieves were moving forward. But Suchi pulled the initiative. Removing a scroll from his pack he prepared to cast a spell. Sleep may not immobilize all of them. He knew what had to be done. Waving his hands he said strange words in Cleric. The spell was cast. The thieves were suddenly stopped in their tracks. They simply couldn't move. Many were locked in posture in mid-stride. Others were just realizing the problem. But all of them had been effectively stopped.

"Move it! Move it! Out now!" Suchi yelled as the others noticed the occurrence. They ran out as fast as they could. When they were a safe distance from their prey Suchi checked for casualties.

"Are you hurt?" he asked Thonolan.

"I'm fine. Thanks to you."

"I knew they would try this some day."

"You know me?"

"I know of you. Thonolan the Brave. I have been watching you. I need you for traps and secret doors."

"I'm yours. I owe you my life!"

"We both owe them." Tralina insisted.

"How long does that spell last? And what is it called?"

"It's called paralysis. And it will last long enough. Hurry, to Avengers Mansion."

Chapter 2

After the company left the waterfront far behind they began to slow their pace. The fighters were running out of energy as all that metal armor slowed them. Suchi stopped on a boardwalk and turned to the others.

"His Lordship has decided on a name for his adventurers. They will be called The Avengers. We will be considered enforcers of the peace."

As they continued The Avengers talked indiscriminately about the interesting adventure they had just experienced. Tralina, the maiden, remained silent.

"I wish we could have slaughtered them all. I despise chaotics."

"At least we had some blood shed."

"There were too many of them. We may not have won against them. Chaotic thieves are fearless and death is subliminal to them. They live to die in battle."

"Still, I would have loved to take as many of them as we could with us."

"Me too!" Roderick stated.

"We have one rule in this group." Suchi lectured. "Avoid violence whenever possible."

"What a drag." Roderick insulted. "I'm a fighter. It's my mission to slay chaotics."

"There will be other times, I assure you!" Suchi promised.

As they talked Suchi led them to the outskirts of the town, the residential area of the city.

"Welcome to Avengers Estates." Suchi welcomed as he approached heavy wooden doors entering through a ten foot high wall of stone. As they passed through the gates Samson noticed that the wall surrounding the property of Lord Eugene Brunweger were made of marble, glossed over and smooth as a newborn. No aspiring thieves could climb the wall or even use a grappling hook. There were no edges for the hook to grasp. As Suchi led them over the well manicured lawn the others looked around in curiosity. On both sides there were buildings which Suchi explained were used as stables. The smell of horse manure permeated the air.

"The master keeps a regimend of horses for riding if we need to go long distances. We will be riding out of town in the morning on horse back. We already have our first mission at hand you know. Now, come with me into the mansion. His Lordship awaits us."

Entering through a pair of furnished doors, The Avengers found themselves in darkness expect for torch light which was used for lighting inside the mansion. Walking into the centre of the hallway they could see that the hallway broke off into two branches. The hallway narrowed up ahead. However, there in the middle of their path stood a round obelisk wall. Suchi motioned to them as he walked around.

"Come with me." Walking around to the opposite side of the obelisk they could see an entrance. The obelisk was hollow. Entering into darkness Suchi pointed out a long pole which rose into the distant roof. A second floor was apparent.

"Can you folks climb?" I'm afraid I haven't a levitating disk put in yet. I can't seem to find the proper spell. Up we go."

Suchi began to climb to pole with Samson right behind him. Tralina was given a boost and Thonolan came behind her. The two fighters followed suit with some difficulty but both made it to the top. When they reached the top each of them jumped off in turn. The dark was still present. Suchi reached in a pouch and opened a scroll. Casting a spell the darkness melted away into light. But this light had no source. The cleric was indeed powerful.

The room included four doors, two on each side. It was obvious these led to other rooms. But inside this enormous room stood a long table with eight comfy wooden chairs. As they turned around in a circle they noticed a fresh supply of ale on the bar at the opposite side of the room. Samson suddenly noticed he was thirsty.

"May I?" Samson asked.

"Make yourselves at home. This is your home now. It will be for a long time."

The Avengers all crowded around the bar and poured mugs of their favorite refreshment. For Tralina Suchi suggested a flask of sweet-tasting wine. As the cups were filled Suchi pointed out a place for each person to sit around the 170' table made of golden marble. It glowed in the magic-conjured light.

"If your are all comfortable I will call this meeting of the Avengers to order. Please state your professional names and affiliations."

"My name is Samson Leader of the Dwarven Territories."

"My name is Tralina of Village Tralesta."

"My name is Thonolan the Brave of Colina."

"My name is Roger Bloodbath of the neutral fighters' guild."

"My name is Roderick Bloodbath of the same affiliation."

"And my name is Suchi Evil Killer, acolyte to Lord Eugene Brunweger, mayor of Colina."

"So, what is our first topic of today?"

"We make up the first adventurers of the Avengers. More may join later but for now we are it." As Suchi talked someone could be heard climbing the pole. When he appeared everyone gasped.

Lord Eugene Brunweger was a monster of a man dressed in elegant robes of torquoise blue. The designs sewn into his robes were to stress is position in the Colina hierarchy. He was six feet in height and grew a goatee and beard the color of Hellhound red. His eyes glowed with the color of crystal. His body and facial features were developed with the look of toning. Every muscle in his body seemed toned to perfection. He was incredibly strong the fighters could tell.

As he turned to the table he noticed the brightness in the room without much surprise. Then he turned to the bar and poured himself a drink before seating himself at the head of the table. Suchi sat next to him and it continued on down the line. He checked out each face in turn and was surprised when he saw the Elven female.

"I did not expect to meet such beauty tonight. What is your name, my dear?" Tralina was exasperated at the man's size. She sputtered once before she replied.

"I am Tralina of Village Tralesta."

"Welcome, my friend. You will be a great asset I believe."

"I must inform you, sir, that I have very little skill."

"You have your magic, I believe."

"I don't even have that. I am an acolyte. My father was teaching me to cast spells but died before he could complete my training."

"You have a magic book?"

"I have only one spell and even that I cannot cast completely."

"Show us." There were murmurs of agreement all around. Complying to the request, Tralina stood. Holding her hands in front of herself, she flexed them in a circle. As she brought her hands together at her waist, a flame of fire began. The flame surged into a ball of fire and a sound of intense heat could be heard. The ball grew, and as she spread her hands forward, the fire suddenly evaporated. She could not sustain the cast completely.

"It is a beginning, at least." the mayor complimented.

"I may not be of much use to your company, I am afraid!"

"Oh contraire. You have your in-born infravision! And I believe, with some training, you may be excellent at hand-to-hand combat."

"I am adequate with the dagger."

"That is another asset! Don't count yourself out totally! I believe I can help you more than you can know. It is advantageous that our first mission is in the Elven Territories. I have a feeling you will benefit from this little venture of ours, my dear. Even if you are not very skilled. Still, someone will have to protect you at all times. You are still very young and there is much danger of attack towards you." Everyone agreed to that observation.

"I will attend her!" Thonolan suggested. "I am skilled."

"However, you will also need protection, my friend! You are an excellent pick pocket but a poor hand-to-hand combatant. No insult meant. Yes, you will both be welcome. Well, back to the matter at hand."

"Thank you, sir." Suchi enthused. "We all have been chosen for different reasons. But our skills are all that have made the decision."

"Samson Leader is a well-known guide and has been all over the principality many times over. He will be our guide to the different areas we must go to. Wherever our missions lead us."

"Thank you for your faith, my lord."

"You are a strong man. Perhaps as strong as myself! You are well-known for your prowess in killing hellhounds! We may meet some on the way. You will be the head guard on our venture."

"The Bloodbaths are known as the best fighters in Colina. First Roger, and now Roderick after him. We will have need of your talents. I am proud to call both of you my counterparts! Now, does everyone know our mission for the morrow?"

"Refresh our memories if you please!" Thonolan suggested.

"Ludwig Wildlife, headman of the village of the same name, has been captured by a human magic user called Andrew. This same magic user has captured the Elven Temple of the village and at last word was trying to get to the village's treasury. He will use everything in his power to make the headman talk. So our mission is obvious. Destroy the magic user and protect the village's treasury. At all costs. The headman's daughter, Sasselia, will join you to the temple once you arrive. She is powerful and will try to destroy Andrew. But she needs your support. Andrew has, at last word, charmed some followers and holds the temple hostage. We will need all the support we can find to conquer the magic user, and you here will have to do. I only hope you are enough to do the job."

"We should make plans to educate Tralina so she is of more help to us. At least get her some more magic."

"That problem will be solved once we arrive at the village. She will be, if they are willing, an acolyte to Sasselia. We can always ask." There was another round of agreement. Tralina smiled at the prospect.

"The horses will be saddled and ready come morn. For now, I suggest you all report to quarters and get some shut-eye. You may sleep in any of twelve rooms. Four up here and eight downstairs. I will guide those sleeping on the main floor. The four up here are adequate if you can move around in the dark. That means we should use those with infravision up here. Any suggestions?"

"Tralina and I will gladly sleep up here."

"Good, I will be here too. You are welcome to have your pick."

"I will sleep downstairs, I believe. But I will certainly bring a flask of ale with me!" Laughter enthused at Samson's suggestion.

"We will also reside downstairs, I believe. My brother and I will spend the night together." Roger suggested.

"Then til morning to you all. I will not be here come the morrow. I am needed at the town office, I am afraid. The misery of being mayor."

Chapter 3

At sunrise the Avengers walked into the stables and as promised the horses were saddled and ready. After eating a scrumptious meal of bread and their favorite drink they were preparing to bid Colina farewell.

Suchi led out each horse as it was chosen. He saved his own mare for last. Her name was Sunshine, as she had a golden fleece on her pelt which darkened at the legs.

"We will need plenty to drink on this journey. And I have already stored sufficient iron rations in each pack. Ration properly. If you run out that's your problem. Samson, you may want wine flasks, but bring water as well. Wine doesn't exactly quench the thirst." Samson nodded compliance.

"Suchi." Tralina called. "Can Thonolan and I ride together?"

"I thought of that so you get Flaky Crust. She's Sunshine's dam and she is older but she's also very strong. She used to be known as a champion. She won several medals at the local fair. His Lordship bought her when she was foaling and he thought she could be a big help."

"Have we got everything?" Roger wondered.

"Roger!"

"Yes?"

"You and Roderick leave your metal armour behind. I have lighter leather you can wear. Metal will only encumber the horses too much!"

"I'll go tell Roderick and we'll change."

"Understood."

The Avengers left through the estate's gates as the sun was in mid-horizon. Riding at a steady pace, they rode the outskirts of the village until they arrived at the city gates.

"Good day, folks!" the gate guard called to them. "Where be your destination this day?"

"We are headed to the Elven Territories to destroy an evil magic user."

"Yes, the filth. I wish they would all die but chaotics seem to pop up in unexpected places."

"Indeed they do." Suchi responded. The gates were opened and soon the riders were out in the plains.

Suchi looked around. Plains and farm land stood in all directions. "Which direction, Samson?" he requested.

"We follow the river heading east two days' distance and turn north. The Elven Territories start on this side of the river which twists to the north toward the Black Mountains."

"Shall we follow the river directly or divert a distance?" Thonolan wondered.

"Makes no difference to me." Samson muttered. "I have plenty of implements. Just keep an eye out for those vicious wolves. I hate hellhounds!"

The Avengers rode til midday, but it was still plainsland as far as the eye could see. A stand of trees stood in the near distance.

"Those trees look like a good place to have lunch. Plenty of shade for the horses and hopefully water."

"The horses don't need water. You can't imagine their endurance."

"Dismount. There's no need to tie the horses. We just let them graze."

As Samson dismounted he scanned the area with his glazen eyes. Not a creature in sight. "It looks safe for now. Just wait until nightfall."

"Hellhounds are night stalkers."

"You can say that again." The adventurers had their lunch and then continued their trek. As they left the river they continued north-east.

"Tomorrow at midday we should come to the forest. That is where we will enter the Elven Territories."

"It's getting dusk. We have to find a place to camp." Suchi suggested.

"Perhaps near this creek. It will give the horses fresh water." Suchi prodded Sunshine to a run, so the others matched his pace.

"Some one has to keep watch. It will be sunlight again in six hours. Three two-hour watches will be sufficient. Thonolan and Tralina can sleep through. I will take first watch. Samson will take second. And one of the fighters will have to be third."

"I'll volunteer." Roger suggested. "I may be old but my sight is as good as ever."

"Done!" Suchi exclaimed. "I'm ordering everyone to get some shut-eye. I'll wake Samson at the chosen time."

The adventurers rode off again a t sunrise as the night had been spent without incident. The rode a distance at a pace and then slowed to ride alongside each other. "How much drink do we have left?"

"I have two full bags of water. Roderick also has two. With your three we have plenty." Roger analyzed.

"And I have two water and half a skin of wine yet. As well as one skin of ale." Samson revealed.

"Drink the water and save the rest for later. We will need drink once we arrive at the village." The others nodded assent.

"It's no use saving the wine. It's half empty. Besides, it's probably full of backwash." The others grimaced at Samson's suggestion.

"The wine's yours then." Suchi compromised. "But the ale is not to be touched."

"If we don't pick up our pace we'll take forever to get there." Suchi suggested. The others responded willingly.

Four hours later the group met the dense forest of the Territories delegated to the Elves. As they rode they slowed their pace to enter the forest with Samson in lead.

There was a well-layed path among the trees which stood some distance through the foliage. But getting to it took some dodging of dense undergrowth. Samson stopped at the path.

"This path leads directly through the Elven Territories to the Black Mountains beyond. If we take it we will have easy access to all the villages of said region."

"Then let's go!" Roderick suggested.

"But we must be careful. This be wolf country. Silver wolves, fir wolves that blend in, and most of all hellhounds!"

"There's plenty of dead wood for fires."

"That would be fine." Samson confirmed. "But in a hole in the ground. A forest like this could burn for days without end. And there's nothing to stop it but heavy rain!"

"Do you want to break for lunch? We can make a small fire."

"It's up to you, boss!" Samson answered.

"Is there a clearing anywhere to set up in?"

"Lots of them! Set up every five to six hours' distance especially for the convenience of travelers. The first one is just up ahead." Samson acknowledged. "This trail is hard to find, you know. I just came upon it the first time by accident. Since then I have followed it all the way to the farthest village where the territories meet the mountains. It's called Village Denvil. Named after its first headman and the name has been passed down the generations as all the villages are named after the first headman."

"Let's break for lunch."

"The clearing is up ahead." Samson insisted.

Twenty minutes' distance took them to the promised clearing which stood off to the right side of the trail. The men set up camp in the small clearing which was only large enough for one camp. No one had used it in a while.

"There are firepits in each clearing, as you may have noticed. We can start a fire if you want!"

"That's awful convenient." Suchi suggested.

"Just common courtesy." Samson assured. "The Elves treat men like kings."

"I noticed."

"I'll go get some kindling." Samson offered. "There's lots in the bushes."

"I'll go get my tinderbox and prepare. I still have to unload Sunshine."

Samson walked to the south of camp knowing he was moderately safe but keeping eyes and ears open all the same. He walked to a distant tree and started breaking low branches. Then he bent over to pick up dead kindling.

No sound except the wind in his face. His ears picked up the breaking of dead wood all the same. Then again two minutes later. He wasn't sure what was stalking him, or if indeed he was being stalked at all. The stalker might have other prey in mind.

Shrugging it off he continued his work. But in his doings he could feel his own anxiety grow. The creature pounced a minute later! In his side vision, Samson had seen the hellhound cub leap! As he did, Samson reached for his dagger!

Chapter 4

Samson yelled in anger as the wolf threw him down to the mossy ground. However, the animal had overestimated his leap. He landed with all fours on Samson, so that he could see the underbelly of the hound. Samson thrust the dagger with all his might.

The aim was good, and the dagger cut deep into the chest cavity. But Samson could tell that he had missed his aim. The heart still pounded, so Samson removed the dagger as the animal backed up.

Hot searing heat from the hellhound's breath could be detected, but Samson wasn't swayed. Grabbing the animal around the neck, he threw it down and turned on top.

Another breath of fire singed his hair, but Samson was enjoying this tremendously! Lifting the dagger high, he plunged it into the chest cavity once more, where he knew the heart was. This time his aim was true. As the six-inch blade bit into the animal there was a sudden gasp of fear. In effort, the animal expelled its last breath.

The rest of the travelers finally arrived. Samson rolled off the animal and yelled his achievement. One more dead hellhound!

Getting to his feet, Samson inspected the slain animal. The first thrust had caught it between two ribs. The second, but more deadly thrust, had penetrated higher up and to the side.

Samson was covered in blood but none of it, gratefully, was his own. This had indeed been a grand fight. The hellhound had been strong but inexperienced. That was Samson's advantage.

"Told you! Hellhounds abound all over this region. This was only a cub... perhaps six months old."

"It's certainly big enough!" Roderick insisted. "I'm glad you were attacked and not me."

"I was alone in the woods. He would never have attacked a group. You see why we must be careful." Samson reached to pick up his kindling and left the carcass for other carnivores to get rid of.

Stepping out of the wood Samson saw for the first time what a mess his armor was in. Blood was soaking into the skin and had created a huge stain. Suchi brought a waterskin and a soft hide to wash off the excess. Nothing could be done about the stain.

As the others built the fire, slowly but efficiently, Samson sat back and smiled his pride. Another success!"

"Hellhounds will live in family groups, but never in actual packs. They are individual loners. In family groups they live together for a short three months and then they separate. During the three month duration the parents teach the young to hunt. Any offspring not walking within five weeks of birth is considered weak and is slain."

"How big of game can they kill?" Roderick considered. "It looks like those beasts can take down anything."

"This one was only half grown. In another six months it would have taken the length of Suchi's height or longer. A single hound can kill a roe deer!"

"How? Deer are fast and very agile."

"Yes, but hellhounds are just as fast at a distance. They are good long distance runners. Now, you ask me how they can kill a deer? Simple. They go for the backs of the legs. That slows them or stops them completely!" Roderick whistled in amazement.

"So hellhounds are a force to be reckoned with!"

"No doubt. But they prefer warm meat. Cooked not raw. And their breath goes a long way to achieving that. Ever wonder how a hellhound breathes fire?"

"Tell me! I would be interested."

"Me too!" Roger assented. "I always wondered."

"It's a long story. About the time I killed my first. It was full-grown and experienced which added to the danger. That was the toughest and most dangerous fight I have ever undertaken. But I won. I slit the throat, you see. When the blood started pouring the animal darted and I don't know where it died."

"But after I killed my second hound I dissected it to see what the vital organs looked like and their general positioning. I still use my knowledge of these vital organs and their positioning to slay my kill. I have killed six hellhounds by penetrating the heart. That is my forte."|

"Anyway, while dissecting, I opened up the stomach. It was full of an acidic substance that I highly respect. I used a tinderbox on it and it burned. It's naturally flammable and I could not recall seeing it anywhere else. I wondered how the animal took liquid acid and produced fire. I opened the throat next. Looking in, I could see an organ in the throat that when stimulated creates a spark. So when the hellhound breathes fire it belches acid and lights it."

"And much like our bile it comes up in mass. The spark lights it, and, depending on how much mass is in the mouth, fire can be produced." Roger concluded. "Perfectly reasonable."

"Yes it is!"

"We better get going if you want to get to the village before dusk. We don't have any time to waste." Suchi demanded. "Let's break camp."

Before they reached the next campsite Suchi thought he could hear the trampling of branches in the bushes. He stopped to listen.

"What's wrong?"

"Shhh." Suchi shushed. "Listen."

The trampling continued a minute later. Something or someone was walking toward them. The sound was indeed coming closer. It was darkening so Samson looked with his infravision.

"It's human. Or at least it stands on two legs!"

"Hello there!" Suchi yelled. "Over here." The man turned toward the path and approached through the undergrowth. He stood at five feet six inches and was definitely Elven. However he didn't come close. He seemed wary of strangers.

"How far is Village Wildlife from here?" Suchi asked. The man just stared at the men like he hadn't heard. "What's your name?" Suchi continued. The Elf only looked more confused. Tralina said something to the man which some of the others could not understand. But Suchi picked up on the hint. She was talking Elven, her own home language. Suchi tried to remember his Elven language. He hadn't used it in a while, he was forced to admit.

"My name is Suchi." he opened again trying his hand at the man's language. Behind me are the rest of my troup. We're looking for Village Wildlife. Are we close?"

"My name is Laurence Wildlife and you are very close. Forgive my misunderstanding. I do not speak the common language very well. I am restricted to Elven I am afraid. That is why I cannot leave these territories. It is a failure on my part."

"Could you lead us there?"

"There is a shortcut through the bushes but you have horses."

"They're better for long distance travel."

"I know. I was checking my traps. Rabbits and hares are rampant in the early morning and late at night as it is now." Samson nodded his head it obvious realization.

"I am familiar with Elven trapping techniques. We must discuss it." The man named Laurence looked surprised to see a dwarf who could talk Elven so well. "I have traveled these parts extensively and know your language well."

"Indeed you do." Laurence confirmed. "Well, we may as well take the long route." He entered into the path and lead ahead of the riders. Those who did not speak the language followed slowly and without question. Soon the path led into the village proper.

"You are here to rescue the headman, I presume!"

"Yes we are. We are here to help!"

"Then you were expected. But not so soon. You are to be complimented on your promptness."

"Thank you, my friend."

"If you give me your horses I will bed them for your stay. There is no need of them until you return to your home."

"You are too kind."

"I am here to serve."

"Understood." In common he told everyone to dismount. As Suchi reached firm ground he turned to see a woman touch him on the arm. The sight of horses had spread the word and everyone was talking in their native tongue. The sight of the Avengers prompted much gossip.

Suchi heard the talk and picked up the gist of conversation but his Elven was rough at best.

"Do you speak common?" he asked to the woman at his side.

"Of course, although most can't."

"We are here to help Sasselia Wildlife." The woman visibly cringed.

"I would not use that name. She prefers to be called simply 'Sassy'."

"Sassy."

"Short for..."

"Sasselia." both voiced.

"She does not like to be named after her full title either. She is very melancholy. Princess Sasselia of Village Wildlife."

"Princess!"

"She is the daughter of the village sorceress. Also known as the headwoman of any village."

"I understand."

"Come. I will bring you before her."

"Thank you."

Sasselia Wildlife was a woman of true character. Her looks were very good as well. Long blonde hair hanging loose around her shoulders, a short but disciplined nose, and a small, roundish chin. Her eyes were a shade of blue. She wore a close fitting tunic of fur. It looked like roe deer hide. Very good at insulating against the cold but very hot in summer. After the group was seated at a fire within the wood shelter they had entered Sasselia Wildlife spoke.

"My name is Sasselia of Village Wildlife." She spat out her full name for convenience. "It is good that you have arrived so promptly. Please introduce yourselves and state your affiliations."

"Suchi Evil Killer, acolyte to Lord Eugene Brunweger and resident of Colina."

"Roger Bloodbath of the Colina neutral fighters' guild."

"Roderick Bloodbath of the same fighters' guild."

"Samson Leader of the Dwarven Territories."|

"Tralina of Village Tralesta."

"Thonolan the Brave of... The Avengers."

"Avengers?" Sassy wondered.

"That is what we call ourselves now. We are a team that Lord Brunweger put together personally."

"Lord Brunweger is a highly respected man. I will take your word that you are the best in your particular area of residence." Sasselia pointed toward Tralina. "She is one of us. I don't understand why one of us is traveling with adventurers from so far away."

"She is one of your kind. As she mentioned she is of Village Tralista."

"That is ten villages away! Why has she gone so far from home?"

"If I may?" Tralina spoke up. Sassy just nodded. "I am from Village Tralista. My father and I were traveling when he grew sick. I think he died of fever. He was very old. I am only an acolyte and would take great pride in receiving training from someone of your stature."

"Noted. Please continue."

"I met Thonolan the Brave after bringing my father to the nearest town for treatment. Unfortunately he could not be saved."

"You are lovers?"

"We have chosen a platonic relationship. At least for now."

"Good choice. I have seen interspecies mating is possible. But there can be difficulties. I suggest you continue in the same way... for now. Meanwhile we have a mission to accomplish."

"Please, if you grant permission, I will speak!" Suchi put in.

"You may."

"It is on Tralina's behalf. We wish to keep her on as a member of the team. However she has very little training in the magical arts. I would be honored if you would consider taking her in as an acolyte, that she is, and training her personally."

"She has no skill?"

"She has a partial spell which she can use but can't cast."

"Show me." Sassy insisted. Tralina responded by casting to her limit.

"You must push your arms out fast and with much force, my dear. The secret in the cast is in the aim as well as the thrust." Sassy demonstrated by flexing her arms out but not casting the fireball itself. "There is still the chant which must be learned. Without that you have no luck which prevents backlash."

"Backlash? You mean I can be hurt by my own cast?"

"Exactly. That is why aim is so important. You must have an object to cast it at. Otherwise it will boomerang back to you!"

"There's so much to learn!"

"Yes, she has the will and the interest. That is enough. I will take her as my most highly respected acolyte. I should have chosen someone long ago!"

"Thank you!" Suchi gratified.

"Acolytes have no place in battle, however. She will stay behind."

"Understood."

"Her boyfriend can keep her company. We will not need a thief on this venture." Thonolan shrugged with assent.

"The dwarf I will need as well as the two swordsmen. I fear there will be bloodshed. Andrew has charmed our own people and I hate to destroy them but that may come."

"How do we get in?" Suchi wondered.

"Andrew has the front door guarded. It will be impenetrable. However, we have a secret entrance that he does not know about. We will use the magic portal."

"Magic portal?" Thonolan wondered.

"It is a passage from one point in space to another. It is instantaneous. It is hidden in the forest near the temple."

"I've heard of them. Very deep magic."

"Too deep for Andrew's good. We will eat now and in the morn we will embark on our mission."

Chapter 5

Sassy Wildlife walked into the shack shared by the Avengers. She appeared anxious and eager to be off. Her worry over her father's welfare showed on her concerned face. She stepped in and gently kicked each of the Avengers in the shins.

"Up, lazy bones all! I've been up and meditating on my spells for two hours already! Today we kick ass!" Samson was up in a flash as the rest followed suit.

"Let's go right away!" Roderick suggested.

"Relax. We have all day! " Sassy responded. "Besides, Suchi has to prepare some scrolls."

"How'd you know that?" Suchi asked.

"Just an assumption. Anyway, you should all grab a light breakfast!"

"What's to eat?" Tralina asked.

"There's plenty of fruit left. And some spare rations if you're still hungry."

"Fruit! Good idea. First thing in the morning is such a good idea." Sassy shook her head in dismay.

"Hungry?" Samson asked as juice ran down Roderick's chin. Suchi started through his supply of scrolls and chose some; leaving others behind.

"I only have one paralysis left. I will leave it here. Also, my darkness is depleted."

"We'll get you replacements." Sassy offered. Suchi abstained from comment. "How long do you need?"

"I'm ready now. I have five dispel magics, two blindness and one magic missile."

"Magic missile? That's a magic user spell. I've never heard of clerics using it."

"It's right here in black and white!" Sassy whistled her delight.

"Everyone finished?" Suchi asked. Everyone nodded assent. "Then lets get out of here. We have some blood to shed!"

"Everyone ready?" the Avengers nodded assent. "There is light in the temple so we won't need those torches. Besides, I have my light spell."

"Me too!" Suchi insisted.

"Follow me."Sassy led the way through the woods using a well-used path. A distance along she turned into the woods, scraping away dry limbs in her path. Soon they approached what appeared to be a stone building which stood in a small clearing.

"I took this pathway so we would not be seen, This is our secret entrance into the temple. It holds our dearest secret; the magic portal." With that she led the way around the back wall and into the building.

The hall inside stood thirty feet wide and opened into two alcoves. Sassy stopped in the middle of the room and turned to her followers.

"This is our most sacred secret of all. If one of us requires easy access in a hurry they are open for use. But they can also be a quick escape mechanism. This is what we call a bi-teleport system. The right is the transporter and the left alcove is the receiver. Each of us, as we go through, will suffer the chills of portal transport. It's cold, but still livable. Also, I'm sure our armor will give us some protection. As you can see," Sassy stated approaching the right alcove, "this appears to be a solid wall. But only to the naked eye. All teleports are obvious to the trained infravision eye. It appears a shade darker of blue than all the other walls. That is how they can be discovered. But only with infravision. Are you men ready?"

"Aii!" everyone yelled.

"Then let's go!" Sassy yelled into the distance. As she prepared a dagger in its sheath she stepped into the wall; her back was the last to vaporize. Suchi and Samson followed with the Bloodbaths behind them.

The transference from one spot to instantaneously appear at another distant location was one of uncertain destiny. One never knew where one might end up with strange portals. But this was a familiar system to Sassy and the others trusted her.

It felt like they were moving via vacuum through a long tunnel. The air was ice cold and almost sucked the air out of t heir lungs. Most people preferred to hold their breath at such a time, Suchi knew. Then finally the vacuum stopped, and they were on the other side of a solid wall made of stone.

First thing Suchi noticed was the absence of light. He could barely make out Sassy's image in front of him. He took out a scroll to cast light but Sassy was faster. Instantly the light was brilliant and visualizing was better by far.

After everyone was in the room they looked around. Everyone was standing in the middle of a twenty foot alcove. The opposite alcove was thirty feet beyond and thirty feet long. They knew it was the teleporter back to where they had come from. To the east, connecting to the two alcoves, was a space thirty by ten feet. A solid wall stood at the far end of the eighty foot entryway into the room.

Sassy tried to budge the wall but it seemed jammed. She asked the fighters to try it. They were unsuccessful. Then Sassy remembered something important. She took Suchi's staff and pushed with it against the top of the wall. A loud grating was heard and Sassy grimaced with fear. What if their entrance had been heard?

The fighters helped lift the secret door to a gap large enough to walk through. Sassy listened carefully. No sound came from the temple except the sound of flickering torches.

Peeking around the door she saw the path was clear. Then she walked out into the centre of a hallway sixty feet long and ten feet wide. There was no sound beyond.

Another hallway stood east to west. Just beyond that lay the entrance to the closet room. Not hearing occupation in there she walked west to the end of the hallway. The others followed her to the end and waited. Sassy suddenly took a start.

"The entrance to the treasury is open!" she exclaimed. "Follow me."

Leading the way across the eighty foot entryway into the temple she approached the open door leading to a stairwell downstairs. She took the stairs two at a time, rushing in her haste, and was probably heard from below.

"What do you mean there are none?" Sassy could hear from the end of the stairwell. "There have to be magic books here too! You're elves!"

"That's what I tried to tell you! Ransacking this place is useless. Our treasury is totally isolated to Elven scrolls."

"I can't even read this!" Andrew yelled in his fury.

"Of course not! It's all written in cleric!"

"What?" Ludwig Wildlife laughed in Andrew's face.

"I knew someone might try this so I converted all the language of the spells to cleric. I hope you harbour no hard feelings." Andrew's answer was to cast a magic missile spell. But Sassy was in presence and cast before him. The man grabbed his eyes in blindness.

While the evil magic user was confused Ludwig Wildlife saw Sassy and knew she was the best protection available. Running, he hid behind her and to the side. Andrew finally succeeded in drawing the blindness from his eyes. Then he got a shock. In the doorway stood an elven maiden. And from her gold rings he determined that she was too powerful for him.

Recovery was fast however. He made a defensive move... invisibility. As Sassy looked for him with infravision she was blinded by a shiny glare. He had used the same trick on her.

Dispelling magic she was forced to move out of the doorway and she felt the invisible magic user slide by. Seconds later she was after him. She may not see him normally but even heat sources could be detected with infravision.

As she ran Sassy saw the thief turn toward the stairs. She was catching up to him when he cast a block in her way. She was forced to cast dispel magic again.

Clearing the way she was soon at the top of the stairs. But he was nowhere in sight. She had forgotten about her infravision which could blind her in torch light.

Thinking of a possible solution she ran toward the secret entrance and was delayed by having to go around ten people who were drawing their weapons. Five followed her but the others had been diverted from her path.

Entering through the open secret door she slammed it shut behind her. Voices of anger were heard beyond even through the thick stone. She had no time to waste however. Running into the far alcove she practically jumped through the teleport.

Once on the other side she ran out and through the forest to the edge of the well-traveled road. Just in time to see Andrew riding horseback toward her.

She did not want to hurt the horse unnecessarily. So she ran into the road and grabbed the holster in a quick grab... dragging her feet behind. She managed to grab the holster tighter. Reaching up she attempted to grab the rider and pull him off the horse. But suddenly metal glistened in the radiating sun. Andrew had unsheathed a dagger.

Grabbing the dagger in his fist he attempted to stab at Sassy's hands. But as he sped into the village proper, people's eyes gaping, she was forced to let go due to lack of blood flow. Her hands were falling asleep.

Falling to the ground in a heap, she picked herself up and cried tears of anger at her failure. He deserved to die! But she had been too late. What a first encounter this would make!

Feeling arms around her Sassy wiped tears away and looked at her charred and bloody hands. The skin had worn away on both palms. She decided to lick up her own excess of blood.

Turning back to the temple she hoped no one else had been hurt. The acolytes had been charmed she knew. They could do plenty of damage if they wanted to. Ignoring the pain in her scarred hands she began to run to the temple.

She attempted to open the heavy gates to the temple. They would not budge. Besides, her hands were feeling worse and the pain was just starting to pierce, now that she had calmed down. What a nightmare! Just then the temple gates flooded open and the Avengers walked out escorting ten acolytes in their wake.

"What happened?" Sassy asked.

"All they had were daggers so we took them hand to hand. As it was I dispelled the charm they were under."

"Excellent!"

"What happened to your hands?" Suchi asked as he noticed the charred and stained palms.

"I tried to keep him from running off. I guess that was stupid, huh?"

"No, but I'd get something on those hands. Looks like the ropes did a number on them."

"Thanks, I will!" Turning away Sassy noticed Tralina behind her.

"What are you doing here?" Sassy asked.

"I heard the excitement and saw him ride off. I thought I'd better check on you." she stated as she applied rags to the blood stains. "Come with me. I can do something for that."

"Can you?"

"Ointment for infection and a proper wrap for your hands. I'm an expert at healing, did you know that? My mother is the medicine woman for our village. I always thought I would apprentice with her. But I haven't had this much fun since I left."

"My beautiful daughter. What a problem you have there. Let's get that looked at properly."

"Hello daddy! What do you think of my performance?"

"I'm just glad you weren't hurt any worse. You certainly gave it your best!"

"But I failed." Sassy stated factually.

"So what?" At least the scrolls are intact."

"True."

"You needn't kill to win. Just get your point across."

"And you've made yours." Sassy stated as she smiled at Ludwig. "What took you so long in there?"

"I had to scout the temple and make sure no artifacts were harmed. But Andrew was in such a rush he never thought of the golden dishware or the gems in the far trunk. All the better for us." Sassy nodded assent as Tralina led her toward the village for proper treatment.

Sassy Wildlife entered into the Avengers' adobe and stood up straight.

"I'm here to announce tonight's feast."

"Feast?"

"We are all going to eat together. The whole village."

"What's the occasion?"

"It's our strategy meeting for destroying Andrew."

"We're still after him?"

"Yes!"

"Why? Just for revenge?"

"Come to the feast and all will be explained."

"We'll be there!" Suchi promised on behalf of all of them.

"We're here." Suchi said as the rest of his troup sat at different places near the spitted quarter of roe deer.

"Gather around. The rest of the village will come soon."

Fifteen minutes later there were twenty adults with babies and older children gathered around the fire which was blazing with the sound of crackling wood.

After everyone was introduced, and translators chosen, the meal began. Sassy cut up huge chunks of deer and placed them on plates made of birch bark. Suchi noticed that most of the maidens wore leather over their breasts and fashionable skirts. Other youngsters in puberty were too young to make much difference.

Ludwig Wildlife started the meal with a choice morsel from the deer's rump. Others dug in after him. Half an hour later drinks were served. Ludwig had offered the village wine which he kept in stock. Samson learned to relish it.

Reaching behind him Suchi revealed a waterbag full of liquid. Opening the top he refilled cups with the ale which Samson had brought with him.

"So, what are you planning, Sassy?" Suchi asked.

"Well, I've made plans to leave the village. Try to find Andrew. But that's a long-term plan."

"Where would you go if you left?"

"Back with you guys to Colina. I've already discussed it with Tralina... I would like to join your band... The Avengers... That would be best for Tralina's sake. I'm sure she would like to be with Thonolan rather than live here."

"You're certainly welcome!" Suchi insisted.

"But I need a pseudonym. Another name to call myself. For the village's safety, not just my own."

"Have you chosen one?"

"Sassy Wildfire comes in handy."

"Sounds good." Suchi complimented.

"I had to change my name too!" Samson suddenly spoke up. When he was tipsy Samson was known to bleed his heart.

"What is your name?" someone asked in Elven.

"I call myself Samson Leader. But my last name is Trellis. Samson of Clan Trellis. Not a bad name, is it?" No one answered him so he just continued. "I wouldn't tell just anyone about my lineage but I can trust you folks here. I am the son of Arwald Trellis, headman of said clan."

"You were born a prince!" Sassy exclaimed.

"Yes, I was. But my mother died a long time ago. I've never forgotten her."

"Who was she?"

"She was headwoman. She was killed while attacking a head woman of another clan. Clan Amrose.

Her name was Amelia, my mother I mean. The other woman I didn't know. The war was started over my betrothed. Amelia Denzil. She didn't want me. She wanted Stanley of Clan Belsil. I didn't blame her. But my father did."

"What happened?"

"My father plotted revenge for the disgrace placed upon my good name. She had insulted Clan Trellis by not choosing her betrothed at birth who had been promised to her. But she didn't want me. I don't blame her!"

"What was so bad about you?"

"I was only seventeen, and she was two years older. She had been promised to Clan Trellis's first-born son. That was me. Now, I've always blamed myself for the war. But I blamed my father more. Him and his clan honor. To hell with it!" Samson remained quiet for the rest of the evening until most of the village had left.

"So you blame yourself for the Dwarven Clan Wars, don't you?" Sassy asked suddenly.

"I should have died with the rest of the clan. But someone chose to protect me. I don't know why. All I know is... I am the last living member of Clan Trellis. Some day I will return and claim my throne."

Sassy put her sensitive hands on Samson's work-hardened own and smiled. She had tears in her eyes.

"A lovely tragedy. And you will regain your rightful place. I can see that!" Samson just shrugged his shoulders.

"So why do you want to go after Andrew so badly?" Suchi brought up.

"Because he steals magic books. His way of gaining power is by stealing it from whoever he can."

"That's what he was after in the treasury." Ludwig offered.

"He looked pissed off pretty badly."

"You would be too if all the spells were written in a foreign language like cleric."

"Cleric?"

"Yes sir. My father taught me the language before he died. Said that knowing other languages was an asset. He knew Cleric, Elven, Thief and Magic User. He'd been a traveler in his young age and learned a lot. I believe he passed on just a fraction of his knowledge."

Samson was out like a light now. Apparently the drink had put him to sleep which was now known to be common with him. After ten more minutes of talk about plans and arrangements for the trip back to Colina the company split up and Samson was left in slumber.

Chapter 6

The companions slept 'til midday despite the long trek ahead of them on their return voyage home. Tralina woke up early, explosions magnifying the ache in her head. Since she was hung over she knew the others would be, too.

As she left the quarters she was conspiring a plan to fix a remedy for all of them. She left the village and walked a distance into the forest where there was damp moss. This is where she would find the strong saturation of the special plant she needed. After digging up the dosage she thought she would need she returned home.

When she got back she ground roots which she had dug up and began a concoction of the plant itself bubbling in water over a small fire. When the water had boiled a while, releasing the essential element within the secreted juices, she added the root powder. Then she let it simmer until the substance was thick and spongy. She knew it would be a strong remedy.

After she had solution enough for all her friends, she walked carefully to their quarters and entered.

Walking in their midst she stirred the medicine carefully and tasted. Bitter and strong. Just the right strength to make the skin clear and dissipate the ache to a minimum. She had learned from her mother to make this potent morning-after drink.

"Wake up, all!" she yelled, knowing that she was just aggravating things. "I've made my mother's special morning after drink. Are the rest of you as hung over as I am? Everyone hung over? Anyone who is say Aii."

"Aii!" everyone yelled back, paying in pain for the effort.

"Take this and drink it. It's bitter, not something to enjoy as an after dinner drink, but it will help to clear your thinking and cleanse the blood. It helps the liver to filter out the poisons. Everyone sit up!" she demanded. Suchi got up with a little effort and pain. The others followed their leader; some with more pain than others.

"God, am I dizzy!" Thonolan confessed with his hands to his ears. Shaking his head cleared his vision. Taking a cup from Tralina he winked his thanks.

Suchi tasted first. His mouth made a frown of distaste. But he knew Tralina and trusted her insight. If she said it would help then they would take her word for it. After he sipped it a few times he got used to the bitterness he tasted.

"Bottoms up, boys." he yelled. "You heard the medicine woman. Guzzle it!"

Everyone emptied the cups in one or two gulps. Suchi gathered the cups as suddenly he felt the effects. His vision cleared and his aching blood vessels subsided. Everyone could feel the remedy working and as this was the first time it was used the effect seemed to be instantaneous. It was certainly giving Tralina's reputation as a medicine woman and healer some added credit.

"Are you as good with fatal wounds as you are at morning after drinks?" Suchi asked.

"I need to gather some medicines today and some sinew. In case I need to stitch anyone up." she suddenly recalled. "Give me until tomorrow morning. It's late to start a long trek, anyway." Suchi appeared to think it over.

"If Sassy is coming with us she needs time to say her goodbyes anyway. IF is the essential word."

"She is coming. I'm not about to stay away from my man any longer than I have to. Besides, Avengers Estates is the perfect training ground. She is coming, and I can attest to that."

"I was just jesting."

"It's not funny!"

"Sorry."

As Tralina exited the Avengers' quarters she saw Sassy Wildlife magnificent astride a reddish-maned horse.

"What are you doing on horseback?" she asked.

"Are we going or do I have to blow my top first?" she asked as Suchi exited into their presence.

"Get off that horse!" he yelled. "We are not going anywhere today. We need to recover from last night and the party. Besides, we can't go half cocked after Andrew. We have to be prepared."

"And let him get to Colina before us? Maybe we can catch him up."

"He's not headed for Colina. I know that for a fact. He has a fortress some where in the prairies. If he showed his head through the city gates it would be lopped off instantly. He's hiding somewhere and until we find out where we can't do a thing. Now get off that horse and be rational. This will be a team effort! The first rule of adventuring. Never fight alone!"

"Yes boss!" Sassy saluted and got off the back of her horse. "Good ol' Red Coat. I'm looking forward to taking her with me."

"Have you even said your goodbyes? Your parents are going to miss you. You won't be a princess anymore. You're a part of the team. Only through team effort can we defeat Andrew. He is one as we are many. Surely he can't take us all. Besides, we have the magic and power. And teamwork. Second rule of adventuring. Work as a team."

"I'm going to groom Sunshine now. It is part of the preparation for the morrow. Then I have to get some added rest. I am afraid recovery from the night before eludes me. Come with me." he suggested as he walked Red Coat back to the stables.

"You can learn a lot from observation. You will be learning from the best. Lord Eugene Brunweger himself. Avengers Estates is the perfect training ground. It is a nice place to meditate and grow! Besides, you have never been in a dungeon. You have to keep your eyes and ears open. You must use your senses and powers to optimal levels. But most of all you must use your mind. Irrational behavior does not help. It only makes things worse."

"I know that." Sassy stated as they entered the stables and she took Red Coat in hand. "But I failed."

Suchi had rewritten several scrolls and was preparing to put his scrolls away. But he was concerned about several spells that had been depleted. He would have to ask the village for copies.

Suddenly Tralina ran into his quarters breathing heavily and whining her breath back into normal perimeters. "Suchi..." she caught her breath and started again. "Sassy wants you at the temple. She says it's fairly urgent."

"I'm coming." Suchi exited and walked out of the village. At a fast pace he approached the temple. Tralina followed beside him.

"She has several scrolls to give you before we leave. She's taking them out of the treasury."

Suchi descended the stairs into the treasury and stood back while admiring Sassy's handiwork.

Scrolls sat in neat piles all around her. She turned her head and smiled.

"I'm doing an inventory of the treasury so I can give you some extra spells. As princess and daughter of the sorceress I am responsible for keeping stock. Here!" Sassy handed him the scrolls on the floor.

"Dispel magic, invisibility, and magic missile to add to your collection. Also darkness, blight, an elven spell of rare use, and telekinesis. Last but not least, a rare but powerful spell, talk with animals." Suchi was taken aback at this last spell.

"Talk with animals? I've never heard of talking to them."

"It's a very rare spell that translates animal language into common. Any of the smarter animals can be affected but not the smaller ones. Generally, the smaller the animal, the harder it is to communicate. But you make sure you learn it properly. And learn to make copies as well. It's written in cleric, but it's older language, so you may need some assistance."

"Thanks." he stated, accepting the scrolls as they were intended. "Do you have a supply of these?" he asked. "I don't want to deplete your treasury."

"That's why I'm doing the inventory. I have to make sure we have an adequate supply. As princess I'm in charge of the treasury. But use them sparingly."

"Understood. Do you need help?"

"Nope. I'm done. Here's your levitating disk. I hope it helps a little. I found it among the treasury. We sure are rich." Suchi nodded as his hands were full of scrolls.

"What now?" he asked.

"Back to the village to dispense of the scrolls. Give me some of those."

"Thanks. How'd you know I needed a levitating disk? I never told you."

"Thonolan told me. So I looked. It's very rare but you can copy it and keep it in reserve. For next time. Besides, as part of the team I have to contribute something.

"Thank you. It's just what we need."

"I'll see you later. Call me if you need help translating the talk with animals."

"Where are you going to be?"

"I'll be around. I just need some fresh air. See you later." With that Sassy left his presence. Suchi shrugged his shoulders and set back to work.

"Sassy!"

"What's up, Tralina?" Sassy asked as she ran to her acolyte.

"Can you start my training right away?" she asked with her hands neatly folded behind her back.

"I guess so. I keep forgetting about me dear student. Come with me. We'll practice your fireball." Tralina couldn't help but smile.

Sassy led her to a practice field which contained several obstacles to master. One of them was a solid brick wall standing five feet wide and six feet high. Both girls stood at a mere five foot one.

"Now, reach out your arms and enclose the fireball. Make sure you aim accurately. Remember the backlash. Get the fire started and at its strongest thrust out your arms like a whip. Fast and fluent. Faster you thrust the more air it takes in. Push and aim at the same time. Like this!" Sassy demonstrated by reaching out in front of her. As she gathered flames into herself the fire began to rage. At its strongest she turned her hands palm out and cast whipping her arms out like two oblong tools. The flames were strong and right on target. The wall burned for five or six seconds and then the fire evaporated.

Tralina gave it a try. As she reached out with her arms she felt the first spark of life. As she slowly drew her arms in the ball of fire began to grow. She went to cast but Sassy shook her head. So Tralina drew more power. Now the ball became more intense than she was used to. It all felt kind of scary to her. When Sassy nodded her head she turned her palms out and copied Sassy's whip. This first attempt was to be the beginning of her growth. The fireball never reached the wall but it had been cast successfully. She was learning, Sassy could see. But she would need practice.

"Continue studying until I call you. I have to tell Suchi about your progress. You are learning fast. All it takes is practice. Remember not to cast until the fire is at its strongest. Every time you use it, before you cast, your palms have to feel the flames. Remember how your palms felt the heat? That's how long you have to wait. Magic has more to do with senses than it has to do with power. Feel the energy you're gathering. Let it flow within and without. Your power is spiritual, of the environment, not of the power of death. Let your surroundings protect you. But most of all feel with your senses. Use of them is essential in different ways for different spells. But each sense will be used in combat. Test the air, see into the darkness, watch for any abnormalities. It could save your life."

That night Sassy exited her father's domain with a heavy heart. She would probably not see the home of her birth for some time; if ever again. She had spent the last two hours of daylight in with her beloved parents and now shook a tear from her eye. It was just dusk but the village was soundless save for the cry of infants in the night. Some day! But it was not quite over yet.

As Sassy entered into the Avengers' presence she knelt and sat cross-legged in front of Suchi. Then she spoke.

"I would like to thank you for your dignity and knowledge. You are t he leader because of your training and experience. I understand that. But my failure with Andrew leaves me with regrets. I should have had him towing the line. If only I'd been smarter, if only I had pulled him off that damn horse, he would be done for. I know that!"

"You can't live with ifs and maybes." Suchi responded. "You saved the village treasury and your own father's life. That is what is important. Not death or conquering for what you believe in. The best sort of fight is 'no fight'. You'll swell with confusion and hatred, sure, but next time it won't be the same. I promise you. He can run but he can't hide. We are the Avengers, the law keepers in this beauty. And as the law keepers we must abide by certain rules. Third rule of adventuring, abide by the law. If you take revenge out of hatred then I will know about it. But in a way you have to pity him." The others nodded agreement. "To steal spells is a bad thing, but we have an advantage. We have our open minds, our senses and the tools of our trades. We will win in the end. Trust me and savour your just rewards after he's gone and dead. It will happen if we use teamwork. And our own solid heads."

"In a cave you're always thinking. Your senses must be at peak performance. You may find some mystifying things in dungeons but you work through them. Logically and physically. Everything you do must be logically analyzed or it may cost you your life. Now, you are tired. Lay down with us and spend your first night among the Avengers. It will be a long journey come morn. So we best get some sleep now. Good night."

Sassy did as Suchi bid and lay down among her new-found friends, wrapping a leather around herself as a blanket. Then she went to sleep. But not forgotten were her regrets and hatred. They were to be cherished and courted. She lived for the day when Andrew lay dead in front of her with a silver dagger through his throat. Some day soon she hoped. Some day very soon.

\Chapter 7

Suchi awoke as dawn came across the glittering horizon. He sensed that the light was just beginning, yet, as he sat up, counting heads, he noticed that someone had left early. Lighting a candle and holding it up, he checked faces. Samson and Sassy were both missing. Thonolan and Tralina lay wrapped in each other's arms, Roger and Roderick were sound asleep and in their dreams, but there was on sign of the other two.

As he got up he wondered where they had gone at such a young hour. The trek needn't start until after breakfast. Leaving the quarters he found Sassy with Ludwig and the sorceress. They were discussing quietly in the new day. A small fire burned off the morning dew.

"Morning to you all!" he greeted. "Sassy, have you seen Samson this morn?"

"He went trapping with Lawrence. You remember him mentioning it as you came?"

"Of course. Please excuse me. I must check on Sunshine."

As Suchi left the warmth of the fire on this chilly morn he walked in the direction of the stables. He found Samson and Lawrence deep in conversation.

"Your friend here knows much about my topic of trapping. It is too bad he cannot stay. I would enjoy his company every day." Lawrence complimented.

"I have already prepared my horse!" Samson passed on. "I am ready when you are."

"Patience my friend. We leave on the hour." Samson nodded confirmation. "Sassy must prepare her horse and I will prepare Flaky Crust. The fighters should be roused by now."

"I will check on their progress." Samson offered. As Suchi saddled Flaky Crust she snorted a breath of recognition and consent. She started prancing in impatience for the journey to begin.

"Patience, my dam. In due time."

Samson found the two fighters roused and eating fresh fruit for breakfast. "You must prepare your horses. Suchi rouses you both." At this the brothers got up and walked out. Samson found Sassy still near the fire.

"It's nearly time. Will you saddle up and prepare?"

"It is already done. I was up before dawn. I will feast with my parents and join you soon."

Roger and Roderick entered the stables and saw their steeds waiting. As they saddled their horses they talked intermittently about the distance they would have to travel back to Colina.

"The journey needn't take long." Suchi interrupted. "Two days' journey either way. It will be nice to get back home. I crave peace and quiet. And the practice yard." Samson finally made an appearance followed by Thonolan and Tralina.

"Your horse is saddled and ready to go. Good ol' Flaky Crust. She is a fine horse."

"Sassy saddled and prepared ages ago. She will join us in due time.

"I noticed she has saddled up. She certainly is efficient."

"Perhaps she saddled night before?"

"Nope. This was done this morn. It would make sense to let the horse rest bareback. Unsaddled."

"Understood."

The Avengers had led the horses out of the stables and awaited nearby while Sassy said her last goodbyes. After hugs and words of affection she jumped on Red Coat and they were off.

Racing along at a full trot they left the Elven Territories behind. For Sassy it might be forever. As she left the Elven Forest behind Suchi could see a tear of salt water enter Sassy's eye. He moved astride and patted her side in a comforting gesture.

"The adventure is just beginning." he promised. Sassy wiped the tear away and smiled back.

"Foolish tears."

"Not foolish. It is natural to cry when first home is left."

"I am a fool for it though. With all this talk of Andrew."

"We will get him! I promise. Some day soon."

The first day was coming toward dusk as the reached the winding river. There they camped aside a fresh source of water. As the night progressed Thonolan grew edgy and weary. As he scouted the area on his watch he thought he saw a shadow in the dark. As it drew near he saw the raging pace of the creature. In a moment the space was crossed and the fir wolf was reaching for his throat. As he fought for his life from teeth and fangs he yelled his fear. Suddenly he felt a burning sensation and the night was lit like day for ten seconds. Now the wolf was off Thonolan as it had rolled with the impact. It was dark again and hard to see. But feeling his way around he felt the dark fur of the wolf beside him. His hands told the story.

Feeling around as he smelled burned flesh, his fingers revealed burn marks on the wolf's coat. It was unbelievable that he had not been hurt. Looking up he suddenly saw a figure step next to him. Tralina had a tear in her eye.

"Rabid! The wolf was rabid!"

"Seems like it. It's lucky Tralina was there."

"But how?" Thonolan wondered.

"Fireball!" Sassy concluded. "It's dangerous but effective. She did the right thing!"

"No one else heard?"

"Everyone heard. Only Tralina was close enough to help."

"I heard you yell. Even in my dreams I know a cry of pain or fear. I could not sleep while you were on watch. I believe I saw the creature before you. With my infravision. I was not fast enough. Fireball takes time to cast. I was lucky not to have killed you, as well." Thonolan nodded his head as he bent and kissed his thanks and relief. Tralina responded with a jump of glorious joy. Her heart began to beat faster and adrenaline began to run. The kiss was long and lingering. When they finally parted it was with regret.

"We'll not get anymore sleep today. Let's suit up and go. It shows on dawn anyway."

The day passed quietly so as they approached Colina they cheered their safe passage. The Avengers arrived at the Colina gates as dusk approached.

"Open up for The Avengers." Suchi insisted. As the gates opened and they rode through the gate guards looked on with envy.

"Those are fine steeds!" the head guard named Walaree exclaimed. "Where is your quarry?"

"The vile thief ran away. If you hear any word in your passings of the magic user named Andrew I want word immediately!" Suchi insisted.

"Understood. I will keep my ears open."

"Thank you."

"Who's the Elven maiden who sits astride the red horse?"

"Sassy Wildfire be her name and I would be courteous if I were you. She be a sorceress."

"Indeed. And who be the other woman with the gent astride the same horse?"

"I am Tralina of Village Tralista. And I am the medicine woman of the band."

"We could use a good healer here in Colina. The one previous was old and died. We have been without for three weeks now. We sent word but nothing as yet."

"If the patients can be brought to Avengers Estates I may be available." Tralina offered with a questioning look at Suchi. He nodded consent.

"It can be arranged."

"Done then. I know of several patients who await in line. I will inform the nurse to come to your quarters. She is also a midwife and has been quite inadequate in her treatment She lacks skill, you see."

"Send her to me, then. If possible, I will teach her my art."

"Fine. I will arrange it personally."

"Thank you."

As the troup moved on they rode easily and soon the Avengers gates loomed ahead of them. Suchi opened them and the others rode through. He walked his horse to the stables. The Avengers unsaddled their horses and lay down fresh hay. Then they entered the mansion and were home.

As the weeks passed by they began training en mas. They learned jousting, fighting with the sword, how to use a bow and arrow, even the crossbow. They also learned how to take a fall as well as hand-to-hand combat. Sassy taught how to throw a dagger accurately. The brothers taught the fighting skills taught only to their guild. Swordplay. Soon Sassy was as good with a short sword as either of the brothers was with the long sword.

Tralina practiced her healing arts as well as learned to cast more spells. Her second spell, for example, became Magic Missile. Her third was levitation although this was harder to control. She also learned paralysis, light and darkness, and blindness which was a variation of darkness when cast at the eyes.

Within two weeks Tralina had a magic book and wrote all her spells in that Magician's diary. As time passed Thonolan and she drew closer tò each other. Some times he helped her with her treatments.

The nurse, named Sandra, taught Tralina about childbirth and in turn Tralina taught about morning-after drinks as well as her other herbal medicines. This was quickly becoming an exchange of information between the two healing women. This created a bond where they were dependent on each other. But Tralina knew that eventually the bond would have to be severed.

Three weeks into the month Tralina delivered a baby boy to a couple of high status. Sandra congratulated Tralina for doing it by herself. Than night Tralina took Sandra upstairs.

"It's so nice to have a successful delivery." Tralina nodded agreement. "What are we doing in this fine room? I have never been up here before."

"Sandra, it's time I set the record straight. In the passed three weeks I have taught you everything I know about medicine. Why? Because I'm not always going to be available. That's where you come in. You're my sub. When I'm away you're it!"

"Yes, I know. And I do appreciate it. But what if something comes up that I don't know about?"

"Then I won't know about it, either." Tralina smiled admittance. "Don't worry. We will cross that bridge when we come to it."

Lord Brunweger appeared regularly and slept his nights at Avengers Estates. One fine day near the end of a month since they arrived home Tralina and Thonolan approached Eugene personally.

"We would like to talk to you for a time." they requested.

"Fine. Upstairs?" They nodded agreement. Eugene entered the obelisk and signaled up. The levitating disk took them upstairs. Entering into the confrence room Eugene Brunweger sat down.

"This is about you both."

"Yes."

"Marriage?" They nodded agreement with an expression of stunned silence. "You two have come a long way in... how long?"

"Six months since we met."

"That long. There's no question you are in love. I recall Suchi saying you kissed when the rabid wolf attacked.

"Yes, indeed."

"And you intend to go on kissing, I am to understand. But why marriage so soon? After all, six months is a very short time to know each other before you take that big step."

"We would prefer intercourse." Thonolan admitted. Eugene just nodded in understanding.

"Why talk to me?"

"You are the mayor of Colina. As that is your office you have qualifications for matrimony."

"But I am not adequate. If you were to elope I might be adequate. But there is no reason for that. Surely the church is the adequate office."

"We do not have access to the religious rights."

"Sure you do. Suchi is registered as a cleric of the Clerics' Church of Antiquities. Surely he is more adequate to perform the ceremony that I am. Let's ask him." As Lord Brunweger rose he stomped his feet in ecstasy. "We will have a festival in honour of your marriage. After the ceremony of course." He approached the disk and went to the lower floor, leaving the couple behind to talk alone.

"You really think it's possible?"

"Lord Brunweger does not lie." Thonolan comforted. Tralina nodded assent.

"I knew marriage was on the way." Suchi stated as he entered on a run. "What about pregnancy? Are you willing to voyage during pregnancy?"

"I never thought of that. What should we do?"

"Stay celebate for now. We will arrange the marriage promptly. But it will take time. I have to arrange a marriage registry and inform the church. Then we must set a date!"

"How soon?" Tralina asked.

"Next month at the latest. It is necessary to prepare properly. Trust me. I will help. I have so much to do! I must start immediately of course. I leave for town on the hour. Promptness is necessary. Trust me it will be soon!"

"Thank you from both of us." Suchi winked an eye and then he was off.

"You two are dismissed. I must also leave for the office. Much work collects on my desk."

"Thank you."

"Just keep your loins together until next month. It's not long. They you two can settle down safe and sound within Colina walls. I am willing to purchase a private home for the both of you. Surely you require total privacy. I will arrange it. After all, you're in good hands with Lord Brunweger." he reminded with a chuckle. "Until then."

Chapter 8

"Open the gates for Permeseus. Open the gates I say!"

"What's up, Permeseus? What's the rush?"

"I have an injured man here."

"You don't say. Let's have a look-see."

"I found him unconscious on my door step. I took him in and laid him on a cot, but I'm no doctor!"

The patient wore a full beard and a magician's robe covering a suit of leather armour. The suit was covered in blood and a hole pierced the side of the armour. Undoing the leather Permeseus revealed a vicious gash in the side of the patient's chest. It looked infectious and burned to the touch with fever. The victim was indeed burning up with fever.

"Open the gates! This man is injured badly!" As the gates opened, Walaree appeared and inspected the wound.

"Pretty serious cut. Looks like a dagger wound. Take him to Avengers Estates. The gates are never locked but it's still a private place, you'll find. Tell them your story."

"Ask for Tralina and state your mission. Then you are free to go home. Charlie, show him the way. I'll give you an escort to get you there promptly."

As Charlie got into the buggy Walaree patted the horse charmingly.

"A fine horse indeed. Be off now. And promptness is necessary. A life hangs in the balance. Adios!"

Permeseus waved the reins and the buggy was gone down the street and into the city. Charlie gave directions as Permeseus raced the buggy at full speed. He didn't know why but he did not want this man to die. As they approached the estate gates a sign loomed in front of them:

IF THIS IS AN EMERGENCY
YOU ARE WELCOME
OTHERWISE
THIS IS PRIVATE PROPERTY
NO TRESPASSING

Charlie got out and opened the gates as Permeseus practically ran past at a fair pace barely slowing down to let him off. Charlie ran through the gates and up to the house as Suchi appeared at the door.

"Charlie, what word? What brings you here at such an ungodly hour?"

"We have an injury. A dagger wound and the patient suffers with fever."

"Let's see."

"He's awake!" Permeseus revealed as Suchi looked down at the wounded man.

"Help me!" the patient whispered under his breath.

"Be still! We will move you gently into the house. Then Tralina will look after you. You are safe here!" The patient nodded comprehension.

"What we need is a backboard." Suchi noted as the rest of the Avengers made an appearance. "Samson, go into the stables and get one of the backboards from the loft. And bring it here!" The patient was soon hauled onto the backboard and everyone pitched in to carry him into the house.

After he was placed on a clean cot in one of the spare rooms Tralina looked at the wound. "Still bleeding. Bring me some alcohol. And dry rags." Sandra complied as Talina prepared her med kit.

"It's a pretty bad gash. But it does not appear to be life threatening. It will have to be stitched up though and cleaned thoroughly. Thonolan, get me some fresh water from the pump. Put it into a fresh container and I will wash the wound."

"Here's the alcohol." Tralina poured a dose of alcohol into the wound then cleaned it with clear water. Then she set to work stitching the gap closed.

"His leather is ruined. Blood-stained and pierced. That was a dagger wound alright. But it was an older blade and it left rust in the wound which made things much worse. But obviously it was thrown from a distance."

"That's for sure. He threw it even though... he is very evil. He must be destroyed."

"Who's he? Who did this to you?"

"The chaotic magic user named Andrew."

"Andrew? Do you know where he's holed up?"

"I came from there. He sensed me... I used invisibility to escape but he knew where I was anyway. I managed to escape, but in the process he threw the blade at me. Unfortunately, I walked into the blade's path. It's lucky it wasn't worse. I remember walking a distance into the prairies holding my ribs tight. When I reached the shelter I just dropped. I guess I passed out. Next thing I knew, I was in bed with a hole in my ribs."

Permeseus entered the room and the patient stared at him with bright eyes. "It's him. He's the one who kept me."

"So what's your side of the story?"

"I found him on my doorstep when I got back from the field. I could tell he was losing a lot of blood. I took him in and let him lay the night. Then this morning I brought him to you."

"I owe you my life." the patient stated. "Many thanks from Lazarus the magician."

"Only doing my duty!"

"You've done more than just your duty." Sassy suggested, moving to the front. "You've brought us word of Andrew and the location of his fortress. We are all in your debt."

"I don't understand."

"The man who did this to Lazarus was Andrew, an evil human magic user who steals magic books to gain power. He must be stopped because his process of stealing is to kill first and steal later."

"That follows." Lazarus offered. "I guess I was lucky. He thought he could charm me. I was unaffected but I played along. Soon I was teaching others how to cast spells. He has several apprentices. As I am a magician by rank I have many spells to offer. And I am also licensed as a teacher. But Andrew demanded more. He wanted me to teach him my art. I tried but learning takes time. He grew impatient and we had several arguments. Soon I knew I must escape."

"But he knew what you planned and went after you."

"I blew my cover. I don't know how."

"There's other ways to know the truth without the obvious. Even if he hadn't suspected, he would have eventually tried to kill you. He's like that. He's evil!" Sassy raised her sword above her head. "Death to Andrew!" Sassy raged her vengeance.

"Amen." Tralina responded.

"Death to Andrew." Sassy repeated.

"Death to Andrew." everyone began to chant. "Death to Andrew. Death to Andrew. Death to Andrew."

To Be Continued

This is the last page. THE LAST PAGE? What happens next? Will they get Andrew? Or will he escape whole again? What about the marriage? Will Thonolan and Tralina really marry? And what about pregnancy? So many questions. Want more? What's next? Find out in The Eight Avengers. Only available through the author.

Epilogue: I hope you enjoyed this book. Please join me in looking forward to volume two which I am planning as we speak. Again, I hope you enjoyed reading this book as much as I enjoyed writing it. Please look forward to The Eight Avengers coming as soon as possible. I thank you for your patronage. And also for your interest.

Yours truly, Marcel Rene Chenard The Author

Book 2 The Eight Avengers Acknowledgements

I would like to thank all the employees and volunteers of Schizophrenia Society of Alberta, Edmonton chapter for their continued cooperation of my endeavors.

I especially thank J. Colin Simpson for his continued support and the access to the office utilities. Due to SSA's support I can self-publish free of charge as I work for them on Fridays and that, Colin explains, makes up for the paper and ink.

I would also like to thank Carolyn Lastiwka for her continued support and suggestions. She suggested the use of Ukrainian names for Colina and Anasthasia as well as the prince's name, Leo Anasthasius. My many thanks fro everyone's continued support.

Your favorite author, Marcel Rene Chenard

Chapter 1

"Death to Andrew!" Sassy exclaimed as she held her short sword once more. "Death to Andrew!" Her smile of glee revealed her bloodlust. Suchi grimaced as he saw this. Walking to her he took the weapon away.

"Patience, my dear. I don't want you acting out of vengeance. We will give Andrew a chance to redeem himself."

"But he's chaotic. He'll never give in or change. He is a menace. Aren't we here to get rid of chaotics?"

"We are here to uphold the law. That is all. If Andrew attacks first or intends serious harm then we will deal with it. But we must watch our step. Rule number four. Never take the law into your own hands."

"What a drag. I want him dead! Look what he did to my poor hands! He threatened me with a dagger! What more do we need?"

"He gave me what could have been a fatal dagger wound. He must be extinguished!" Lazarus put in.

"I'll talk to his Lordship and ask. But don't get too bloodthirsty before we are even ready. We must prepare. I will call a conference of the Avengers as soon as Lazarus is recovered enough. Meanwhile, we must continue to train. We need optimum skills to get rid of Andrew. We must be the best!"

"And we will be, I promise you!" Eugene Brunweger was standing in the door with a huge smile on his face. "I require an update. How are our trainees coming along?"

"They are learning fast. Thonolan is more efficient and prominent with the dagger! Sassy fights as well with the short sword as either of the fighters. Tralina has learned quickly and carefully. She has a total of nine spells in her magic book!"

"I would like to see them perform. Then I will decide if they are ready. Meanwhile, how is our new patient doing?"

"He is sore but alive. I have stitched up his wound and prepared a compress to alleviate the infection caused by the rusty blade. Really! Andrew should take better care of his weapons." Tralina extrapulated. "But if you really want to kill someone, this is the slowest method I know!" Lazarus laughed at her comment.

"Indeed, death is slow to come!" He laughed until he cried out in anguish and pain. "My lord, he must be destroyed. Andrew is a danger to society. He must be extinguished!"

"And he will die. I have put out a warrant for his arrest. Dead or alive he must be captured and dealt with. Promptly. He will die, I guarantee it. And Sassy will have the honors. He's dealt us all enough hardship. Now it's payback time!"

Sassy grabbed her sword back from Samson and raised it once more. "Death to Andrew."

Eugene approached Suchi outside of the patient's chamber. "How soon will Lazarus be able to move?" he asked.

"Tralina says he needs two weeks' bed rest. And she plans to stay at his bedside every minute."

"Tell her she cannot be miscounted for her treatment. But make sure she gets enough sleep. Lazarus is in no danger. She needn't serve him hand and foot."

"I have a feeling she wants to serve him because he is our only connection to Andrew."

"I think his rank has more to do with it. He is a magician after all. And indeed he would be a superior asset."

"So you intend to ask him to become one of us."

"He is too priceless an asset. A real magician is indeed too priceless to lose." Suchi nodded his head. "He is the eighth Avenger. That will be our maximum. I have run a check on Lazarus the Magician. He is the real thing. He is registered with the local magic users' guild at his claimed rank. Master Edward, the sorcerer who runs the guild, told me himself and gives me a good reference. Indeed, Lazarus is too precious to lose. Lord only knows what his magic book looks like." With that Eugene walked out the door to the training yard and was gone.

As the days passed, Lazarus healed. Five days after his arrival all that was left of his wound was a scar and some stitches. This was the day that Tralina would remove them. Lord Brunweger called a conference in the afternoon and everyone knew that Lazarus was going to be honored at this conference.

As the morning passed Tralina finally removed the stitches painlessly and had Lazarus stand on his own two feet. Tralina had kept a patch on his chest to secure the wound, but a mere five days after his arrival he was already on his feet. Tralina helped him to the upper floor and sat him at the head of the table next to his Lordship. Suchi sat in the next seat.

"This conference of the Avengers is called to order. I hereby acknowledge the presence of all concerned."

"We are here to honor Lazarus the Magician. He has not only come into our presence but he has also brought us word of Andrew's location. For this we all honor you and we ask that you remain as one of us. We would like to include you as an Avenger. How do you speak?"

"I am honored and glad to be a member of such a fine troup of men and women. I have two fellow magic users to work with and a troup of fighting men that are the best. I would be proud to call myself an Avenger."

"Then this makes up the entire troup of men for our team. Two elves, a human magic user, a thief, two fighters, a cleric and a dwarf. Perfect combination. As there are now eight of you this will be the entire team. From now on I deem that this troup will be known as The Eight Avengers. After all, we are the strongest force in the principality if not beyond. I thank you all for your cooperation. Now, we will enjoy a drink and toast to Lazarus. Dear friend, we thank you! To Lazarus!"

"To Lazarus!"

Another five days passed and Lazarus was harping to get out of bed. Tralina insisted he take it easy at first. She helped hold him up as he took his first steps. When she let go of him she saw that he could hold his own. As he stood he took careful steps. He realized, however, that it would take time to get his balance back. Sitting in bed so long had not only impaired his balance but it also caused bed marks and his back was sore. Tralina reassured him that everything would be normal after a while.

On the sixth day Lazarus decided it was time to get back on his own two feet. So he got up and moving to a chair, dressed in fresh robe and new leather armor which Suchi had provided.

After he was done he slowly stepped through the door. While there he met Sassy.

"You're on your feet. It's about time I would think! Would you like to practice with me? I'm reviewing a few of my spells. Interested?"

"Yes, I am. But I may be slow at first."

"You're doing fine. One step at a time, after all."

"Thank you... outside?"

"Of course... out in the training yard, I believe."

"Onward, then."

Sassy led out to the practice yard which was located behind the mansion. The estate wall continued around the back yard of the estate which closed to the practice yard.

"So what do you want to start with?"

"I'd really like to see your fireball." Lazarus insisted.

"Certainly. We installed a brick wall just last week. I'll gladly show you mine. Then you can show me yours."

Sassy cast her fireball and whipped it directly at the wall. It hit and charred the centre... then died.

"Very nice, indeed. But can you make your weapon curve?"

"Curve?"

"You may recall that the fireball boomerangs?" Sassy nodded understandingly. "Well, if you can calculate the ball's trajectory and diameter you can curve it into an object."

"What's trajectory and diameter? I've never heard of those terms."

"Trajectory is direction and diameter is the distance through the fireball. Let me show you!" Lazarus walked a distance away from the wall and stood facing Sassy toward the north. The wall stood east to west.

"Watch!" Lazarus estimated the calculations. Then he cast his fireball northward. As the weapon moved further away it started to curve into the westerly direction and hit the wall directly dead centre. The fireball had curved into its desired target. Sassy jumped with joy at the sight.

"How did you do that?" she asked.

"The weapon always moves in a circular direction. Depending on the strength of the whip it will be driven so far straight and then naturally curve. The tricky part is knowing the strength of the whip and the distance you want before the curve. That has to be calculated in your head. It takes practice, but it's possible."

"I've never heard of curving your fireball or tried it. Can I learn?"

"It's very dangerous. A skill only for masters. Or magicians."

"I can understand that! What a talent!"

"What do you think of the merits of magic missile?"

"Merits?"

"Does it work every time? Can a missile cast by magic kill every time?"

"It's supposed to."

"I think the death has to be believed in. If someone believes he will die, he will. Also, I think magic missile is circumstantial. Not everyone dies of that spell."

"I haven't had that much experience with the spell combat-wise. I've never used it to try to kill. I just use it in defense."

"As is meant, but it's too bad you can't curve magic missile."

"That would be interesting. Then you could attack a moving target."

"That's my thought as well. But I never want to see you curve a fireball. Never! It's dangerous and if your calculations are off you can be seriously injured. Promise me! Straight lines only!"

"I promise. And thank you for showing me a few of your secrets."

"I'm an Avenger now, remember? We are a team. As I've never been part of a wholesome team before. It feels good!"

"Well, you're certainly welcome among us. It seems to me you could have killed Andrew a long time ago!" Lazarus just shrugged.

"I know how much you hate him. I share your feelings. But a lawful person never offends. Defense is the key."

"I'm more neutral, I guess. That's probably why I hate him so much."

"He has no morals... and your morals are drastically offended. That is what is behind your bloodlust. You are morally offended by him."

"You seem to understand me better than myself."

"I'm certain of that. After all, I have intuition galore. Unfortunately, it wasn't working when I first met Andrew."

"He is complicated in his own way, isn't he? A simple intellect, but a complicated mind."

"As beginners are."

"He is only a beginner."

"But lucky to get away from skill such as yours."

"My hands are my proof. I should have been smarter. I should have cast magic missile first."

"Would it have killed him? That's the question. The spell depends greatly on belief."

"You think so?"

"Magic works because of belief in general."

"I think it's more dependent on skill. That's always been my motto!"
"Perhaps. Everyone has his own beliefs about magic. But it's not guaranteed!"
"Like paralysis. I've seen that fail often enough."
"Understood."
"What's going on here?" Eugene Brunweger came into notice, "Casting more spells?"
"Did you know that Lazarus can curve a fireball?"
"I heard. Master Edward gives a fine reference, my friend. Feel free to use the yard at any time."
"I think I will meditate on my spells for a time. Please help me back to my room, Sassy."
"Of course. You still need help?"
"Just interested in continuing this interesting conversation a while longer."
"Thank you! I enjoy your company, too!"

Chapter 2

On the day that denoted a month since Lazarus came to the Avengers, Lord Brunweger announced the contests. The Avengers would be tested for their might.

I was a sunny summer day and the mansion was too dark and gloomy for the Avengers' good. Tralina was outside with Sassy mastering her magic missile. Roger and Roderick were standing on the edge of the practice yard practicing their swordsmanship. As everyone seemed to enjoy the sunshine, Eugene decided to take advantage of it. He called Lazarus and Samson outside along with Suchi and he called for silence.

"It has been a month since Lazarus joined us. Your skills have been tested and somewhat improved. Now display your skills to your best. I will be here to judge. Sasselia, you will get your short sword. Bring all your weapons and skills to fusion. You will fight for yourselves against your own mates. And make it good! At this moment all your lives are at stake! From each other! You will fight to subdue. And conquer. To the tests."

Sassy went in to get her short sword. Thonolan displayed his fashionable selection of silver daggers. Tralina would stand by as medical help as well as contest with her spells. Lazarus would display his curving fireball. It would indeed be a fine day for mutual combat.

He contesting began with the sword battle between Sassy and both fighters. As they stood in face-off position the swords were placed above and in front, joining in a union of peace. But soon, Sassy knew, she would be tested to her limits with two to one.

Sassy moved first. Swinging around in a circle she attacked while neither of the brothers suspected. But both swords clashed with her own and the attack was deflected. Then Roger attacked with a forward thrust. However, that was neutralized easily. Roderick came from behind as Roger attacked but Sassy had side vision and turned just as Roger's sword was repelled.

The sound of clashing swords continued for a full ten minutes as neither side could get an advantage. The Sassy suddenly changed her methods. Running into Roger's blade she kept hers held high. Roger was watching the blade and didn't notice Sassy's next move. Suddenly Roger found himself on the ground with Sassy's blade lying at his neck. The sneaky woman could think on her feet. She had used a foot sweep to take him down by tripping him up at his feet and jumping to his jugular.

Meanwhile, Roderick had also been taken by surprise to see his brother subdued. Sassy immediately relinquished that Roger was bested, observing all sides. Which is hard to do while a sword is after you. But she had proven to be Roger's superior. She had cheated in a sense. She had combined hand-to-hand with swordplay. She knew it had been the only way she could take him down.

Now she faced Roderick, and although he was the weaker hand, he was the more witty. He knew to think while he fought and used sight more than his brother. But only a little more.

The swords moved. Sassy thrust forward and the blades slid along each other. Then Roderick stepped back and thrust from below. Sassy repelled him easily. Roderick called out to her as they fought. "Well, be aware that I will be doubly careful."

"I'm counting on you giving me a good fight!" Sassy replied. The swords clashed and repelled each other once more. Now Sassy looked like she was going berserker.

"Now, now, now. Don't lose your temper. It will destroy your concentration." Unfortunately, he was the one losing the concentration and repelled Sassy's blade badly at the next thrust.

"Who's the weaker?" Sassy asked.

"You are!" Roderick insisted as he crossed his blade in front and got past her sword. He went straight for her throat. However, Sassy saw the move and stepped back two paces as she slid the sword into view. They stood facing each other if face off once more. Then Eugene intervened.

"This is enough for me. It looks like Roderick and Sassy are evenly matched."

"I had her in submission!" Roderick complained.

"Yes, but she stepped back and parried! That is a fair move in any swordplay. Come, my friend, let us cheer up and call it a draw. Agreed?" Sassy and Roderick looked into each other's eyes and both smiled. Shaking hands, they both had a better appreciation for the other's abilities.

"Next is the dagger throw tourney." They drew straws and Tralina ended up Thonolan's competitor. This pleased her endlessly. She could test her skill against her lover's. Sassy would cast off against Lazarus. Then the two winners would cast off against each other. Another draw was cast for which team would go first. Tralina drew against Sassy and Tralina won the draw.

A target had been cast against the stone wall and the first team took out their daggers. Tralina chose to use silver daggers for their sharpness and versatility. Thonolan also used silver daggers but kept regulars in reserve. Each would throw three daggers. Then the target would be judged. Tralina started.

Holding her first dagger by the blade end, she aimed carefully and thrust her hand out, releasing the dagger. It flew and rotated on its pivot as it was thrown. Hitting the target, it sunk into the cardboard and landed in the outside circle far from the bull's eye. However, this was just a warm up. Thonolan threw next and it was thrown from the hilt. It landed in the second circle closer to the bull's eye. Both of them retrieved their daggers and the contest truly began.

Tralina threw her first dagger after waiting for accuracy. It landed in the third of four circles around the bull's eye. It was a good throw and pretty accurate.

Thonolan then threw his first dagger. He threw it by the hilt as he was not sure he could succeed if the twirled the pivot. This dagger landed in the second circle, so Tralina's dagger stood the successor.

Leaving both daggers alone they prepared for the second thrust. As Tralina aimed, she winked an acknowledgement to Thonolan. Then she began to concentrate on her task. The second dagger was also thrown with a pivot and landed in the third circle closer to the bull's eye. Thonolan then had his turn. Aiming carefully, he threw and it landed in the fourth circle very close to the bull's eye. So, two daggers had been thrown by each. This would be the deciding thrust.

Tralina held her third dagger by the blade and lowered her head to her neck to relax. Then she aimed and threw.

This dagger pivoted on its axis and landed within the bull's eye a little to the outside. She smiled with triumph. Thonolan winked back in glee. To beat her he would have to land his blade closer within the bull's eye. He dare not mess this one up. Grabbing his last dagger from his belt he lowered his hand and concentrated. Then he cast. But, unfortunately for him, it landed in the fourth circle just outside of Tralina's dagger. Tralina jumped with triumph and pleasure. All the same, Thononan hugged her and shook her hand. Thonolan was only pleased to see her skill and accuracy, even if it had been a contest.

Sassy stepped up next with Lazarus. The two previous contestants retrieved their daggers and the target was made fresh. Sassy would throw first as she was female.

Testing the air with a wet finger she judged that there would be no friction between wind and dagger. She announced that she would not need a warm up. Lazarus consented to starting the contest without further adue.

Sassy paced herself and raised her first blade. Grabbing it by the hilt, she aimed... but the hilt felt too heavy. The blade had a thinner structure and the weight felt better. So she decided not to be outdone by Tralina. Carefully, she aimed and pivoted its axis as it flew. This first cast landed inside the fourth circle outside of the bull's eye. Lazarus raised an eye brow in surprise and appreciation, showing he was impressed. Lazarus prepared his first cast.

Reaching at his waist, he removed a dagger from his makeshift belt, a belt that had not been made only for him, but for each one of the Avengers. Taking careful aim he threw this blade from the hilt. He admitted that his style may not be as good but his accuracy may be a means for contest. As it flew it landed just outside of the fourth circle, outside of Sassy's thrusted blade. Sassy prepared for her second thrust.

Aiming carefully, she shot from the blade once more and succeeded in getting the dagger within the fourth circle closer to the bull's eye, closer than her first cast.

Lazarus smiled his appreciation and took his stance. This second dagger landed inside the fourth circle closer to the bull's eye than either of Sassy's casts. Sassy meant to get serious at this last cast. Her third cast landed within the bull's eye just on the outside edge. Lazarus nodded in confirmation.

"Beat that!" Sassy challenged. Lazarus smiled back her challenge and began. Standing with feet apart he took careful aim. This took the better part of thirty seconds. But everyone gawked as he hit the bull's eye closer in. Lazarus had won this heat.

The final heat would be between Lazarus and Tralina. Sassy smiled her surprise and hugged Lazarus in acceptance. All blades were removed and the last contestants set up. However, Tralina was sure she could never compare to Lazaus' skill. But she thought she would try.

Tralina started with a warm up throw. It landed in the third circle deeply imbedded. Lazarus' throw landed in the fourth circle but not far from the last blade. Then the final contest began.

Removing their blades, they paced themselves off and prepared. Tralina took the start. Aiming carefully, she thrust her dagger with her usual style. It landed in the third circle on the outside. Lazarus took her place and prepared his first cast. He knew he could throw better and with more accuracy if he didn't pivot his throw. His blade landed in the fourth circle close to the bull's eye. Tralina gulped in aggravation and a small bit of contempt.

Aiming her second throw, she took her time. The blade landed in the fourth circle outside of Lazarus' blade. So the magician had better stamina and a better cast despite his old-fashionedness.

Lazarus cast his second throw. It landed outside of Tralina's closest dagger. It had been a bad shot and he cringed with regret. But there was still one more cast for each of them.

Tralina aimed carefully, and once again she took her time. Her aim was good, and it landed within the bull's eye on the outside edge. Lazarus knew this would be a challenge.

Aiming carefully he cast the last throw of the day. It landed inside the bull's eye, yes, but so close to Tralina's last throw that they needed an examination. Close check revealed that Lazarus' shot was the winner. By an eighth of an inch. The contestants all shook hands and could now relax. The gaming was over.

Finally one more event was to come. Lazarus demonstrated his curving fireball. Then everyone sat down against a wall of the mansion to soak in the sun.

"I have an announcement to make." Eugene announced. "You have all demonstrated excellent skill and mastery. Therefore, first thing tomorrow, we will have a planning committee for our trip after Andrew. And then we will be off! Death to Andrew!" he yelled. Everyone repeated those words with cheer.

Chapter 3

Toward the evening of the same day Eugene noticed that it was still fairly early and the troup would do better to have the planning meeting tonight and leave first thing come the morn. So he found Suchi and told him to spread the word. The meeting would be held immediately. Meanwhile, supper would be served on the second floor. And not just rations. Home made Roast Beef! With all the stuffings, of course!

"Tomorrow morning we leave on our venture."

"Just keep your weapons and spells handy."

"No kidding! This may be tougher than we thought!"

"Nothing's easy!"

"Especially this venture!"

"Good luck is all I have to say!"

"We will encounter some rust monsters! I know of at least one."

"There goes my suit of metal armor. I was going to wear it to see Andrew."

"No metal."

"Understood!" Everyone nodded assent. Just then the meal was served. Everyone fell silent as they ate, contemplating the work ahead of them.

As the silence grew Tralina began to feel edgy. After everyone had eaten their fill and the roast was taken away she could stand it no more.

"How will we travel?" she asked.

"Horses?"

"I don't think so. We could walk!"

"How far?"

"Five hours distance to the north-west."

"Can we walk that far?"

"Horses wouldn't be too convenient. Especially if we stayed there more than one day."

"Then we walk. Good thing is the five hours is in walking distance. Not horse distance. It won't take long."

"Do we walk all the way? I'm likely to get tired at half-way point."

"We can always take a break." Suchi reassured.

"Then we wait until the next day to attack!"

"Looks like it."

"Can we carry all our supplies with us?"

"We could and must!"

"Bring some short bows, a grappling hook and some torches. As well as your side weapons."

"That should do."

"When do we leave?"

"Sun up?"

"I think," Eugene opened, "that we should leave just before dawn. We could march to the gates and be outside in the prairies by sun-up. So let us go to bed and rise early. It will be a long day tomorrow."

"Maybe we could get a ride part way?"

"Perhaps. But we must wait for morn to find out. Report to your quarters and get some sleep." At this the men and women separated to their individual places and slept.

The Eight Avengers arrived at the city gates at sun-up just as a horse and buggy was going out. Hailing the driver Suchi asked where he was going.

"North-west two hours' distance."

"On our way. Give us a lift?"

"Hop in. Lots of room." So the troup rode as hoped for the two hours' distance.

"How far will we have to walk after this ride?"

"Probably two hours."

"At the most." Lazarus put in.

The men sat back to enjoy the noises of the open prairies. The wheels riding through tough grasses and on rough roads, the sound of birds in the trees and flying overhead, the noise of running water at a stream where they stopped to water the horses.

When the buggy turned into a farm yard everyone grimaced. This was the end of their comfortable ride. Now they walk.

"End of the line folks. Dismount and have a great day."

"Which way?" Suchi asked as he stretched his cramped legs.

"North-west further. Towards the Black Mountains. More north than west actually. Which way's west?"

"This way!" Samson assured as he pointed passed the farm house. "Past the dugout and to the north." So the troup of fighting men went around the dugout and continued in a northerly direction.

"See the wall ahead?" Suchi opened. "That must be our destination."

"Where? I don't see it." Tralina insisted.

"To the north-east." They started walking in that direction.

"I see it! Please, tell me it's not a mirage."

"It's not." Lazarus assured. "This is our destination, alright! See the walls to the east? We go around them and enter from the eastern wall. The gates are there!"

Finally the men and women arrived at the fortress gates. Only the inner doors were standing. The outer doors lay on the ground twenty feet from the gates.

"There's holes in these walls!"

"Let us take a look!"

"I agree. We need a look-see." Suchi suggested. "Get the grappling hook." The hook was thrown over and latched to the top of the wall and Sassy climbed up to the opening.

"I count twelve of them right now! Standing guard at the gate and marching like soldiers. We should be able to take them by surprise."

Meanwhile Tralina was applying her levitation spell to the strewn door which had been lying on the ground. As she moved the door to the north she discovered a hole which had been dug under the door. Placing the door down gently she walked up to inspect the opening.

Suddenly a tentacle-like appendage grabbed her by the ankle and started to pull her to her back. Everyone turned as she was knocked out of breath. Then they saw her freeze.

Jumping down from the wall, Sassy immediately ran to her rescue. The creature, whatever it was, was pulling Talina into the hole. But Sassy took her dagger and cut the appendage which held the medicine woman. Then she dragged Tralina a safe distance.

Roger approached the hole and looked in. The creature was a lump of round flesh with six appendages which reached out to grab him. He swiftly cut the creature's legs off at the base. When he stabbed the creature's fat base, it moved once and then died. Reaching down he pulled the creature from its home. Then he looked in.

Bones from skeletal remains covered most of the bottom. But digging through, Roger found a brown leather bag from one of the victims. Inside he found two stones made of an element called onyx. Digging further he found an old silver dagger which still glittered. After being in the earth for so long it was amazing that it was still burnished.

"How's Tralina?" Suchi asked.

"Paralyzed. It's only temporary. We will have to wait until the poison wears off."

"How long?"

"No way of knowing. Maybe two hours or perhaps all day!"

"Set up camp! We go no further until she recovers."

"Yes sir!" Suchi went to inspect the creature.

"What is it?" he asked.

"Carrion crawler. I've only seen one other." Roger informed.

"Dead?"

"And rendered harmless." Roger assured.

"Get rid of it."

"Gladly!" Roger picked up the creature and threw it into the hole. Then he began to fill it in.

Sassy went back to the wall and took another look in. "They seem to change their numbers but the most I see is twelve in the courtyard. Sometimes less."

"That's good to know." Suchi suggested. "Maybe we can think of a strategy of taking them while they have fewer numbers."

"Is Tralina getting any better?"

"She's not moving at all!" Thonolan informed. "Her flesh is ice cold."

"That's normal with shock." Samson put in. "Carrion crawlers are mean. They paralyze a person and then eat them alive."

"Ouch. Please, don't mention that. That hurts!"

"I'd rather face ten full-grown hellhounds by myself than one of those beasts. Very dangerous." Thonolan nodded agreement.

"Well, I guess we wait."

"For how long?" Thonolan asked.

"Forever if necessary." Suchi suggested. "We cannot afford to lose her."

"I can't either." Thonolan insisted.

It was well after dark before anyone noticed any movement. Thonolan sat beside his mate all night long waiting for her awareness. When she finally came out of it, it was with an indrawn breath and a cry. Rolling on her side she lay in the recovery position for a full ten minutes. Then she looked up.

Thonolan was watching over her and she reached out to grab his hand. Crawling up she sat kneed and hugged him in thanks.

"How are you feeling?"

"Better. I want to thank you for your company. I knew you were there all through. I could see and hear everything but I couldn't move. It was petrifying."

"In the morning, since Tralina is better, we attack."

"And good luck to us all!" Samson blessed.

Chapter 4

Come the morn Sassy awoke early. Rising, she walked to Suchi and shook him.

"If we take them while they sleep they should be easy to take!" Suchi nodded agreement. Pausing at each person, he awoke his followers. Then they began preparations. Climbing the rope Sassy paused at the opening in the wall.

"They're still asleep!" she informed. The others followed her to the top.

"Can you jump eight feet?" Suchi asked as he reached the top after her.

"Just bend your knees."

"Maybe a roll would help!" Sassy nodded agreement. Jumping silently as possible she reached the ground and tucked into a roll. The others followed. As they all reached the ground they walked around silently. The goblins still slept. Picking out eight goblins strewn around the courtyard they yelled a warning. The goblins awoke... too late. They were cut down before they could raise a weapon.

The last four goblins attacked the nearest opponent. Sassy slashed and cut, lacerating a goblin at the shoulder, and crossed, slitting his throat. Thonolan chose a dagger as one approached. He took the creature by behind in a back stab. Two down and two to go! The last two were taken in a jiffy. Sassy looked around. The Eight Avengers were woundless and victorious. Sassy yelled her triumph. Suchi called for silence.

Stealthily they approached the mansion doors which stood at forty feet. Suchi called for positions. Sassy was in front. Then Samson and Suchi. Then Thonolan, in the middle, Lazarus behind him and Tralina watching the back. The fighters took position just inside the doors.

The room stood eighty feet by fifty feet and was closed in except for one entranceway to the west. This led to another room eighty by forty feet which contained three doors leading in all directions.

"Which one do we take?" Sassy asked as she lit a torch.

"Lazarus?" He only shook his shoulders.

"West?" Everyone approached the door straight ahead. As they checked it they found that this door was locked. Sassy called Thonolan.

"Pick the lock!" He pulled out his tools and within two minutes the door opened.

As the room was entered beautiful music was heard. Samson, Suchi and Tralina covered their ears.

"What's the noise?" Tralina asked.

"Noise? It's beautiful!" Sassy insisted. Walking in she saw a table set in golden plates with a fireplace against the far wall. In front of the fireplace stood three creatures which sang the beautiful music. Samson entered and pulled Sassy away.

Noticing the golden silverware and plates, Sassy shrugged of Samson's arm and approached the table. As she touched the plates it sent up a plew of yellow smoke. Sassy sneezed. Then she coughed. Suddenly she fell to the floor... apparently unconscious. Samson pulled Sassy out of the room and closed the door. He then listened for breathing. It was slow but steady.

"What happened?" Suchi asked.

"She inhaled some kind of powder. It may have been poison." Tralina took over and checked Sassy's condition.

"She's not poisoned. It's just knockout gas."

"What's the purpose?"

"Those creatures were horrid! That sound!"

"I think the music was a lure. But it sounded like noise to some and music to others."

"The poison must be in the creatures themselves!"

"I've never seen anything like them!" Just then Sassy awoke. Shaking her head she rose.

"What's going on?"

"Go get the fighters." Suchi told Talina. They arrived promptly.

"Take position in front of Tralina."

"Right or left?" Samson asked.

"Left. Let's get going, Sassy, you feel well?"

"I'm fine."

"Onward then."

This door led to a hallway forty feet across. It continued twenty feet wide into the distance. Across the other twenty stood a stairwell going down. Sassy pointed the way. As Tralina entered last, she let the door go, and it slammed.

Suddenly five more goblins appeared from an open doorway. Battle ensued. Roger disected a head dislocating a shoulder in the process. Roderick ran one through the gut. Sassy cast blindness in a third. As it walked blind, she gutted it with a silver dagger Tralina took the last two with fireball. Blood and bodies were all that remained of the goblins. The amputated head rolled into Samson's view. He merely kicked it out of his path.

Suchi instructed the team to break into parts and search the rest of whatever was possible before exiting into the stairs. The groups parted and returned soon afterward. One group reported a creature in the adjoining room. Suchi led in for an inspection. As the group entered they noticed that it was pitch black. Sassy cast light into the room.

The creature was reddish in colour and completely covered in slime. Its only motivation was its bottom which somehow slid, pushing in on itself as it moved.

"Rust monster!" Lazarus pointed out. "After any metal it can find. I told you!" The Avengers moved out of the room and back into the hallway. Tralina examined her fireball's handiwork in the next room. The burned skin and clothes were revealing. The clothes, torn shreds of dirty material, were charred to nothing. Skin and cooked flesh was all that was left of the two goblins. The room was otherwise empty.

"What now?" Sassy asked.

"We must clean out this main floor before we continue. We need a secure means of escape!" Suchi put in. "We still have one door to explore." Sassy nodded the lead back out the door and to the next. Opening the door they ended up in a similar hallway entering into two rooms and a second stairwell. As they listened they could hear someone snoring from the left room.

This room was dimly lit with torches and four goblins were strewn on the floor. A flask of foul-smelling booze was apparent. The goblins were sleeping off their drink. The troup had a conference and four Avengers slit the sleeping goblins' throats. As they gagged on their own blood the Avengers continued to the next room. Hearing faint throws from the goblins, the troup noticed darkness once more. Suchi cast light with a scroll.

In the middle of the room stood a figure all wrapped in rags of old cloth. It was a mummified undead. Suchi raised his talisman to turn it but Tralina was faster and cast a fireball.

As the creature was hit, the rags caught fire. It went running in panic as it burned. Soon the creature was burning fluently. After a while it stopped walking in panic and dropped. It continued to burn until it was a chunk of black char.

"That's the second fireball you've used. Surely you have no more!" Suchi suggested. Tralina just winked back.

"You're good with that fireball!" Sassy complimented.

"I know." was all Tralina responded.

"Let's take a break!" Suchi instructed. "You folks stay here and I'll do some scouting. This floor should be barren."

As Suchi left the others he went into the next room. The goblins were now swimming in their own blood. And very dead. Next he checked the closets in both rooms. Vacant.

Exiting through the door he entered the opposite end of the mansion. Both closets and rooms were vacant. Then he took a slight risk. Entering into the left stairwell he went down.

The lower floor was very well lit with torch light every twenty feet. To the left stood a wall and to the right stood thirty feet of open space. The opposite wall stood in the distance.

Going back up the stairs Suchi found himself in darkness. The light spell in the room had deminished. He went back to the group safely.

"Let's gather these corpses up in one messy pile. Someone used to live here and I want to show respect." So the mess was soon cleaned up.

"Now what?" Sassy notioned.

"I want one fighter to watch the stairwell. The rest of us need time to rest and meditate."

So Tralina reviewed her fireball, just for good measure. Suchi rewrote his light spell. And everyone took a general recovery period.

Suchi finally announced the move downstairs. Sassy closed her magic book and rose as Suchi kicked Thonolan awake. Then they prepared for the next phase of their venture. Roger came in from the stairwell and announced that everything was deathly quiet.

"It gives me the creeps!" Taking their normal positions in rank the Avengers started down the stairs.

At the bottom Lazarus stepped out of line and led the troup across the hallway to two doors across from each other. He explained that these led into living quarters.

"These are my quarters." he stated as he chose the door to the right. "We will be safe here." Without further caution he entered the room. As he stepped in, however, he saw that another magic user had taken residence. The stranger stood, staring at Lazarus, with a wild look in his eyes.

"Lazarus the Magician. So it has finally come."

"What has?" Lazarus asked.

"Your death!" The magic user waved his hands and a missile appeared at his head. "Die!" the stranger yelled.

Suddenly, Roderick jumped in front of Lazarus and took the blunt of the missile. He fell and lay in a prone position.

"Die, die, die!" the magic user yelled as he stepped over Roderick's prone form and attacked Lazarus with a dagger. The magician responded in kind as both magic users grappled over the same dagger. But it continued its inexplicable descent to Lazarus' jugular. With an effort he pushed the dagger up to head height. Obviously he was the stronger opponent.

Suddenly Lazarus' insane opponent gaped with open mouth. Thonolan could be seen holding the insane man from the back. The magic user collapsed to the floor as Thonolan tore his dagger out of the flesh and wiped it clean.

"I had it under control!" Lazarus argued.

"He was obviously chaotic. Besides, it's better to put him out of his misery. He was a man out of control."

"With berzerker rage."

"How's Roderick?" Lazarus asked as he walked away form the confrontation.

"He's dead!" Tralina announced. "He jumped right into that magic missile."

"Now we know just how effective magic missile really is!"

"Let me in!" Suchi insisted. "It may already be too late!"

"Too late for what?" Sassy asked. Suchi smiled with glee before he responded.

"Ressurection." he answered. Bending over Roderick he turned the body onto its back.

Retrieving a scroll, he waved his hands over the corpse. Then he used both hands, moving them slowly from head to toe. As they moved a field of energy appeared, engulfing the body. Then Suchi went to the centre of the field and pushed with both hands forcefully. Roderick's body jumped with the impact. No result. So he died it again. Still no response. "One more time!" Suchi yelled. There was an audible breath. Then Roderick opened his eyes and the field evaporated. Lazarus took a place at Roderick's head

"My hero!" he thanked.

"Don't try to talk! Save all your strength to heal! Welcome back from the dead."

"I've never seen that spell before! Ressurrection!" Sassy wondered.

"It's only useful directly after death. It doesn't always work. Luckily the force field formed stable. Now he needs twenty-four hours' complete bed rest. He is very weak but alive. His strength will return after he has rested."

"Who will watch him?"

Lazarus got up and walked to an opposite door. It was apparently locked. This confused Lazarus so he took a minute to check for a key. The obvious place was the magic user's robe.

As he opened the door sound came to his ears. The rattling of chains. Looking in he saw that his own five apprentices were shackled and chained to the walls. They looked tired and hungry as ugly bruises were forming where the shackles clung tightly.

"Lazarus! For goodness' sake, you're alive."

"Indeed. I am alive and in good health!"

"Why did you return?" a person with a shaggy appearance asked.

"I will explain everything. First let me remove these chains."

"What about Felonius? He will argue with this! He has held us here ever since the rebellion."

"You rebelled?"

"Andrew could not charm us. When we saw him bringing in magic books, lots of them, we began to ask where he got them. He never answered. Then when we guessed the answers he had us locked up and chained. We were listed for execution."

"That won't happen anymore! You did the right thing in resisting. Felonius is dead and everything is about to change. Soon Andrew will be dead too!"

"At your hand?"|

"At all our hands. I come with a band of adventurers called the Avengers. There are eight of us here. Come now and you will meet them."

"Adventurers? They would dare to invade Andrew's fortress?"

"Daring and proud of it."

"Then they must indeed be powerful."

"So powerful that they brought one of their own back from the dead." At this Lazarus got blank expressions. "Come, and I will show you!" He led them out into the other room.

"These are my apprentices. They were chained and left to starve for rebelling. We can trust these people. They will gladly look after Roderick while we search for Andrew."

"What happened?" one of the apprentices inquired.

"Magic missile!" Roger replied. "He was laying here dead when I got in. He is very weak but he will regain his strength!"

"We will watch him. I'm Selonia by the way. My friends are called Arthur, Marion, Nana and the Elven woman is Maria. We are Lazarus' faithful apprentices."

"We are also faithful!" Roger replied. "We are The Eight Avengers. He's my brother Roderick Vincent Bloodbath and I'm Roger. Suchi, Sassy, Thonolan, Tralina and Lazarus. We are the Avengers and after this I will tear Andrew to pieces."

"Then God-speed. Don't worry, Roderick is in good hands." Roger felt assured.

Chapter 5

The seven Avengers who were left to fight exited into the hallway and approached the opposite door.

"Who's quarters are these?" Suchi asked.

"Permesius lives here," Lazarus explained, "but Andrew visits this room pretty often."

"Then we investigate!" Roger suggested.

"This room may be secured!" Lazarus suggested.

"Then we be careful!" Suchi insisted.

As Sassy opened the door someone could be heard pacing inside. When the door was open Sassy saw the person who lived here. Rations were stacked on one side while a table and chairs with dirty utensils stood in the centre of the room. Beyond could be seen another adjoining room 70 feet beyond. The others pushed her from behind in an effort to get through and she almost stumbled.

Suddenly Sassy felt exhausted when she had rested just hours earlier. She lay down and fell fast asleep.

"Who are you people? Why are you here?" the occupant accused. "Andrew will be through here any time. Leave before he kills us all!"

"Permesius, calm down. We are seeking Andrew to cut him down. We have come to restitute this fortress!"

"Lazarus? You're alive?"

"I have come back with these people to rescue some magic books. Do you know where he resides?"

"Andrew?"

"The same."

"He resides nowhere and everywhere. I do not know where he goes but sometimes he goes into that room and never comes out. Then I see him come in again from your route but he never leaves through that door. He goes in and does not come back out."

"What's in there?"

"My sleeping quarters of course! But he always insists he go in alone. And when I go after him he's gone! Vanished! It's magic, I tell you! He's a winged falcon or something. He just vanishes."

"Can we inspect your quarters?"

"Of course!"

"We need Sassy! What did you do to her?"

"The lady? She's just asleep. Sleep spell, you understand."

"Let me take care of this!" Tralina suggested. "I'm the doctor of the group."

"Cast dispel magic. That will wake her."

"How about you check that room and leave the patient to me."

"As you wish." Suchi acknowledged. "Men, at arms!" Permesius lead the way into the adjoining room.

The room they entered had an inner room with a door that opened outward. After close inspection it appeared that the room within a room was only a construction. Otherwise the outer room was bare. Suchi walked in and noticed that the only furniture in the inner room was a bed against a wall.

"He comes in here does he?"

"At first I thought he wanted to sleep but when I went to wake him he wasn't here."

"Are you sure he never left any other way?"

"He comes in but never leaves this room by normal means!" Suchi smoothed his chin and did some quick thinking.

"He comes in but never exits. That means that this room has one or more secret exits. I would not put it past him! Check the walls for secret doors!' Suchi asked of Thonolan. "Samson, you give him a hand! Something does not connect here. Have you ever seen his exit?"

"He always insists I stay out of the room when he comes. I don't know about secret doors but it's always possible."

"We need a conference." Suchi insisted as they walked back out toward Sassy who appeared fully recovered from her unexpected nap. "When you two are done I want to call a conference. Something's amiss here, and we are going to find it!" Sassy nodded her head and smiled. "Enough beauty sleep?"

Twenty minutes went by before the searchers gave up. "There are no secret doors anywhere in the outer room." Thonolan reported. "I can attest to that."

"None?" Thonolan just shook a negative.

"Then where does he have it?"

"The inner room may be the key." Sassy suggested. "Not many constructions are made that way."

"It is uncommon! Then we search the inner sanctum." Sassy nodded agreement as she led the way.

"A bed against a wall!" Thonolan noticed. "Can't get out of the wrong side that way."

"Or can you?" Tralina suggested in inspiration. Then she blew out all the torches in the bedroom. "See it? Do you see it?"

"See what?" Suchi asked.

"Infravision! This is a magic portal! Take a look!" Sassy, Tralina and Samson all saw what was obviously a change of temperature at the wall the bed protected. "This is the way to Andew's quarters, I know it!"

"Then we attack!"

"Death to Andrew!" everyone yelled.

"In position." Suchi commanded.

Going through the magic portal was much like any other including the Elven portal. The breath could be knocked out of you and the coldest chills. But they arrived promptly at their destination. As they walked around they noticed that they were in a 70 foot alcove that was brightly lit. As they looked out they could see Andrew himself, ready at stance, and he currently had the upper hand. Casting a spell he motioned. The Avengers instantly went to draw weapons but only Lazarus and Sassy were successful. Turning these two looked at their fellows. The others had been efficiently immobilized by paralysis. Before Sassy could react Andrew suddenly disappeared from sight. Sassy reacted by blowing out all the lights with one cast of darkness.

"You can't hide Andrew! I have my infravision!" Sassy roared. "You are dead already." she threatened. She saw Andrew head toward her and then detour toward the paralyzed victims. But Lazarus used levitation to move them out of his reach. Sassy headed him off and since Andrew was blind as bat and helpless in the dark of a closed in shelter, she cast dispel magic and Andrew appeared beside her. Andrew drew a blade as he cast light with one hand. Sassy responded by casting another spell. This was mirror image, a spell that duplicates your form in several mirror images. As the images appeared, the Sassies started circling Andrew as he attempted to find solid flesh with his dagger. But before he could find paydirt Sassy grabbed him from behind, the images disappeared, and Andrew found himself very efficiently subdued in a head lock.

"Yippi kai ai, mother fucker." Sassy yelled as she slowly and painfully slit his jugular open. "Die a painful death, slowly mulching on your own blood!" Then she laughed. "So much for the human magic user who steals magic books!"

Suchi found he could move again. Soon everyone else was back to normal, too.

"Congratulations, you have one dead magic user to bring back to town!" Sassy nodded as he washed her hands of Andrew's blood.

"Search the room! Andrew kept his magic books somewhere."

"I have some information to relay!" Samson offered.

"This room is built like a T. I estimate that the room may be bigger than it seems."

"Secret doors?"

"Or magic portals."

"Check on it." Samson found two invisible entrances to the far sides of the alcove. Inside each room they found trunks full of stolen magic books.

"Any way to return these to their rightful owners?" Suchi asked as the trifled through the trunks.

"Not really, unless one labels a magic book, but that's not recommended."

"Then what to do?"

"Give them to the Archives of Magic. They could sure use the extra spells!"

"Arrange for the return trip to Colina! We'll take the body with us to town and hand him over to the undertaker. Give him a proper burial." Sassy agreed.

"Despite his weakness he was a worthy opponent. I'm sure he died a quick but painful death."

"Quick but painful? You are so cruel!"

"You don't mess around with the likes of me and get away with it!"

Samson volunteered to make a skid and arrange Andrew on it for the trip home. Sassy took time to search magic books to see what spells were available. Later she would share her results.

The Avengers had all returned to the room where Roderick and the acolytes were waiting. Roderick himself felt quite recovered though still not himself. Sassy approached Suchi to report her findings.

"I have searched all the magic books in Andrew's possession. I find that they contain 1st level spells and some second. Invisibility, Levitation, Blindness and Darkness, some fireballs. But the combination of defense and offense indicates that Andrew, a 1st level magic user, would not be able to use most of those spells. Why collect spells if you can't cast them?"

"What do you think?"

"I think Andrew was collecting those spells for a reason. Maybe for someone else to use. But improper use of spells can be disastrous! One must be trained properly in safe use."

"Then he was working for someone, someone who wanted to collect all the spells that he/she could get her hands on."

"But who?"

"Any ideas?"

"None! Even if we had questioned Andrew he would not have responded. Maybe we will never know the whole truth!"

"Understood!"

"How long before we can leave? This place gives me the creeps. Too many bad memories."

"Samson and Roger are chasing a series of gelatinous cubes to pass the time."

"Gelatinous cubes are harmless?"

"Absolutely. They just slide down the halls cleaning and picking. They melt when fired and sometimes you can find things once they are melted."

"And rust monsters? I saw several in the halls and byways."

"Harmless unless you wear metal. They are really harmless."

"I found it! I found it!" someone was heard from the hall. "I found it." Lazarus entered the room with a scroll in his hand. "I found it."

"Found what?"

"The scroll of this fortress. Completely mapped in detail. We have the answers. I have been hoping to find a map and this one is complete!"

"Where did you find it?"

"There was a false bottom in the right room in Andrew's quarters. I looked through each chest and I found this!"

"Is there a signature or any label of ID?"

"Only the initials S.W."

"S. W. Those are my initials." Sassy recalled.

"Coincidence! Whoever created this map simply has the same initials."

"Maybe the map was made by the person who hired Andrew."

"Who's initials are also S.W."

"This is queer. It is almost as if I drew this map."

"No one else we know has these initials."

"Show us the map." Lazarus opened the scroll and laid it down on the floor. Roderick slid close to take part.

"The centre is a massive labyrinth! Rust monsters an gelatinous cubes also appear."

"But look here. This section is a home for five goblins."

"We could take them out."

"Anything worth investigating in the labyrinth?"

"It's just vacant halls encircling each other."

"Very confusing."

"You guys confronted Andrew here?" Roderick inquired.

"Yes. You can see the false walls leading to the treasury."

"Then how did you guys get out of Andrew's quarters?"

"We went out the same way we came in... only this time we found ourselves in the hallway. We had to walk here."

"Here's where you ended up. And unless I'm mistaken that same exit is also the only way into the labyrinth."

"Very complex architecture. Whoever built this place was a very good architect."

"With the initials S.W. and an excellent knowledge of magic portals."

"Good enough to create several portals all over the place and to different destinations.

"This is too important to ignore."

Chapter 6

Three days went by before Roderick was well enough to travel. During that time the Avengers scouted the fortress but found no more encounters. Therefore, the fortress was announced secure.

"We leave as soon as Samson gets back with the horses we sent him to Colina to procure."

"Which should be any time now."

"Samson's here with the horses! And Lord Brunweger is come with him."

"Talk about the devil!"

"He is coming to congratulate us."

Fifteen minutes passed before Mayor Brunweger entered the room. First thing he did was shake hands with each of the successful Avengers.

"You did it!"

"He gave us a lot of resistance."

"But we won!"

"It was a risk. Sassy and Lazarus were the only ones not immobilized by paralysis."

"I understand that. But it's either all or nothing."

"I slit his throat." Sassy boasted.

"Where is the corpse?"

"In the other room. It's probably deteriorated to a stinking stench."

"We actually have two bodies... an enraged acolyte shot Roderick down with magic missile. We barely took him down."

"You took risks, but now it's over."

"Not for me it isn't." Sassy insisted.

"Andrew is dead... all's right with the world!"

"We found this map... it's drawn by someone with Sassy's initials. And we also know that Andrew had an accomplice."

"So what next, then?"

"I would like to look in the village archives and try to find some answers. Who is S.W. and who built this place as well as all these magic portals. Whoever it is will be a trial to deal with."

"I'll start a search right away." Eugene promised. "Meanwhile, we have a return trip to make."

"I have arranged for the matrimony." Suchi informed Tralina and Thonolan. "It will be performed as soon as we can get back to Colina. Everyone's invited. You two will be husband and wife at nightfall." Tralina and Thonolan both shook Suchi's hand and then Thonolan took Tralina into a huge bear hug.

"So you have your vengeance. Andrew is dead and the Elven Territories are secure once more."

"I won't be able to rest until I find all the answers to my current questions."

"I know. But it's over and done with. When we get back to Avengers Estates I am prescribing total rest. You hear that everybody? Our job is done. From now on we can relax and hopefully the peace will last."

"Everyone," Suchi announced, "I acknowledge the joining of Thonolan and Tralina in holy matrimony. They will be joining at dusk today. Everyone is welcome. The ceremony is at the Church of Antiquities in Colina at dusk today." Everyone clapped congratulations and whistled their glee. The noise continued for five minutes.

Sassy approached Selonia who had been sitting aside the whole time with the rest of Lazarus' acolytes and knelt by her side.

"I would like to train Maria as my acolyte. She could learn a lot from a good teacher."

"She is not very knowledgeable and I know that Lazarus hasn't had time to train her properly. She would learn faster from you because she is a little bashful."

"That will change."

"I know it."

"Lazarus, what will you do with your acolytes?"

"Maria will study under Sassy. Learning from a fellow maiden would make her much more comfortable. The others are all human. They can go to the guild where I learned my skills."

Roderick walked over to Eugene to relay his readiness.

"I am feeling much stronger today. Is it okay if I ride home on a horse?"

"It certainly is."

"Thank you."

"Ladies and gentlemen, it is time we left this fortress to rest. Peace be to its walls. Onward to Colina."

It was just reaching dusk when the party arrived at the gates. The town already knew about the death of Andrew. Everyone would gossip and create tales about the adventure for weeks to come.

"Open for the mayor." As the gates opened there was a sudden hush of silence. Tens and hundreds of villagers had gathered at the gates for a slight look at the corpse. But what they noticed more was the look of tiredness in each traveler's face. Men were unshaven, the women were scraggly and many bore bruises on their wrists from chains and there were tears of relief to be free.

The cavalry stopped inside the gates and the mayor gave each citizen a good look at the downed magicians. An hour later they moved on to the centre of town. The city labourers would bring the dead to their graves.

After everyone got settled at the estate and cleaned up there was held a general meeting to discuss the passed adventures while Suchi arranged for the church and service. But soon Suchi was back to get the fiances.

"It is a short walk." Suchi announced. "Just follow me." The upstairs emptied fast and the Estates stood barren once more.

At the church Suchi instructed the lovers to stand together at the pew. The people shuffled quietly in and found their seats.

"We are gathered here to join this man and this woman in holy matrimony. Who will witness this event?"

"I will!" Lord Brunweger yelled out.

"So noted. Tralina, Thonolan, do you both swear to be to each other and only the other as long as you both shall live?"

"We do!"

"Then trade the sharing of gifts." At this point Thonolan and Tralina faced each other as Sassy walked up holding a case between her hands. Tralina opened it and inside was found a pure platinum dagger of rich furnish and shine.

"This belonged to my father. It's pure platinum. It was forged by my great grandfather and was passed down to me. You will give it to our first born son." Thonolan took the knife out of the case to inspect it... and then there was a scream heard that echoed throughout the church. Suchi turned toward the screamer but was suddenly blocked by two thieves who had entered silently along with several other thieves. Suchi thought of drawing sword but did not like to draw in a holy sanctuary. He decided to try to go between them but they refused to move.

"How dare you dessicrate this sanctuary!" he yelled. "What do you want here?"

"We want you. The men your fighters killed were like our brothers. Now it's payback time!"

Suchi finally recognized these men as the chaotic thieves who had tried to kill the groom perhaps a month ago. His face grew red and this time he did draw sword. The fight was short but furious. Soon after it began the thieves began to retreat. Suchi fought them all the way to the door. Then there was silence once more. Suchi held his sword in his hand and looked around. Now it was too quiet. Then he recalled Sassy's scream. Turning he looked in her direction. But she was not to be found at her seat. Or anywhere else. Thonolan strutted to Suchi and leaned on him.

"Sassy's been kidnapped!" he announced. "She's gone! She was taken by the thieves. She's been kidnapped!" Thonolan cried open in grief for their loss. Sassy's loss and the ruination of the best day of his life. Once again the thieves had won.

To be concluded.

MAYDAY! MAYDAY! DISASTER! DISASTER! What a disaster that was. Why would anyone take Sassy? What a ROTTEN DISASTER that was! Attacked right in the middle of a holy ceremony. And by the SAME TROUBLE MAKERS AS BEFORE. SACRILEGE! EVIL! EVIL! Now's the time to panic. Who was Sassy's attacker and why would she work with these miserable thieves? What has happened to her? Is she dead? What can be done?

Book 3 The Wildlife Campaign Introduction

Welcome to The Wildlife Campaign. Inside, find the most intriguing and exciting fantasy novel ever written. The Campaign itself is intriguing enough, but throw in a Centaur or two, and a speaking sword, and you have fantasy galore.

I would like to thank, once more, Carolyn Lastiwka for her explorations and thoughts on what I would include in my novel. I would also like to thank Schizophrenia Society once more for their resources. I look forward to finishing this exciting book and I hope you enjoy reading it as much as I do writing it. The Author, Marcel Rene Chenard

Chapter 1

Samantha Littleton stopped at a solid wall outlining the border of the city of Colina. She had not used the city gates to gain entry into the city and she certainly did not intend to use the gates to make good her escape.

Turning, she watched as the thieves marched in double file with a levitating disk between them on which slept the unconscious body of Sasselia Wildlife. She led them on toward a solid wall. Stepping up, she went through the wall which gave way to let her pass. As she disappeared the thieves began to panic. But remembering their orders, they followed.

As the thieves arrived outside the Colina walls they could see that Samantha had readied a horse and buggy. Using a levitation spell she lifted the disk, which held the unconscious body, onto the back.

Returning to the thieves, she handed each of them a bag of platinum pieces. Then she sent them back within Colina walls.

After the last one disappeared she waved her hands and prepared a spell, speaking archaic words. Slowly the wall in front of her began to glow, and a huge vortex appeared, as the magic portal was irrevocably removed back to non-existence. Leaving behind not a single trace of her passing.

"Report!" Eugene Brunweger insisted, as the informant appeared.

"The city gates have been quiet all night! No one has come in or gone out all night. There has not been any disturbance."

"Thank you. You may go." Eugene gratified. "No one has used the gates so they left by other means."

"They could still be in the city." Tralina suggested. Eugene just shrugged that possibility away.

"Whoever kidnapped Sassy had got in, so we assume the same is the way out."

"Then how?"

"Perhaps Eugene is right and the kidnappers are already gone. With the proper spells, one could make a temporary entrance and erase all traces within seconds. Whoever gained entrance erased all signs before she left the city limits."

"At least we have a good description of the assailant thanks to Thonolan's astute observation."

"I have that sketch now, sir!" Thonolan informed as he handed Eugene the outline of an Elven maiden with black hair and gray-blue eyes. Her nose looked small and pudgy, her ears were extra-emphasized, and she looked to be middle-age in Elven standards.

"Thank you. I'll have this posted and sent around. Anyone with information will know who to watch out for."

"Now what?" Thonolan asked. "Sassy could be dead by now."

"I don't think so. Whoever the kidnapper is will ask for a ransom. Meanwhile we go to bed and rest. If such a thing is possible on such a night as this. In the morning we leave at first light for Village Wildlife and inform Ludwig. He may be mad at us, but there is nothing else we can do here."

Two days and nights later Samantha Littleton drove her horse and buggy into the outskirts of Village Littleton. She had pushed the horse for two whole days and nights, but soon they could rest as long as they needed. Casting invisibility, she vaporized the whole horse and buggy from sight. Then she drove into the village without sight or sound. When she reached the caverns entering into the Black Mountains she released the levitating disk from the buggy and pushed it toward the caverns. Leaving the disk at the entrance she went back, and dispelled magic, revealing the buggy. As she entered the caverns she pushed the disk and its occupant in front of her, leaving the horse and buggy where it stood for everyone to see. She was tired and worn, but soon it would be okay. She was home.

The Avengers had spent a quiet trip to Village Wildlife, and no one had slept a wink throughout the three day trip. They had stopped regularly to water the horses and rest, but not one of them could sleep well. They could still envision their embarassment that Sassy had been taken from in their very midst. There had been no real rush, Eugene knew, and he and Suchi both hesitated to meet the headman of Village Wildlife and undergo his wrath.

On this, the third day of their journey, they reached the Elven forest and trotted slowly along the well-known path through the Elven Territories. There was no real rush, but both leaders knew in their hearts that they wanted to delay the meeting as long as possible. But when, by midday, they arrived at the outskirts of Village Wildlife, their grim faces suddenly grew grimmer. As the horses entered the village there were stares and cheers of happiness as children skittled out of the way.

"Suchi Evil Killer, what on earth are you doing here? And why the so grim faces? Is Sassy with you?"

Suchi just dismounted and asked that Ludwig Wildlife be told that they had arrived.

"What's wrong?"

"Sassy's been kidnapped!" Finally the whole town knew the truth. The word would spread like wildfire.

"Surely it's merely rumor. It can't be true! Who would do such a thing?" The whole town was angry that such would be done to their future sorceress. "Surely it is not possible."

Just then the man of Suchi's summons walked up alongside the horses as the men dismounted. Suchi kneeled at Ludwig's feet and paid regret.

"Sassy's been kidnapped from right under our noses!" he yelled his confession.

"Indeed!" was Ludwig's only response.

"Surely you heard?"

"I heard from my informant in Colina as much as a day ago. And some of the details. But I want the truth of what really happened. Get up and come with me to my quarters!"

"We have brought our master, Lord Eugene Brunweger, for he is the one who protects and nurtures all of us."

"My Lord." Ludwig bowed but Eugene slated it aside.

"It is my honour to bow to you my friend! After all, I owe you a debt."

"Come, we will all talk in private."

Ludwig led the way to his hut and bowed as the entered, and the others followed. Once inside he gestured to the fire.

"Now, I understand that Sassy was captured by a fellow Elven of one of our villages. I have sent word asking for any possibilities but I have not heard back yet."

"We have a sketch of the maiden in question."

"I must see it." Suchi pulled out a piece of parchment and handed it over.

"This is the maiden you saw capture Sassy?"

"Thonolan's the only one who actually saw her. But I trust his devotion."

"Then we must deal with Village Littleton on this matter!"

"You know who she is?"

"Samantha Littleton, wicked witch that she is! She is Sassy's only known enemy."

"Why would she hate Sassy?"

"Sassy is the heir to head sorceress of the entire Territories. Sam always thought the honour belonged to her. But she refused to learn white magic and embraced black. She did not understand the need for defense from backlash. Without the white you have virtually no control over the blackness of the Underworld."

"The Underworld?"

"It's theorized that when beings die their spirits go to the underworld. Like a purgatory. Only the vile or evil beings never or seldom leave the underworld. When they do it's up to the surface they come to spread chaos once more."

"What do we do now?" Suchi asked. Just then a runner entered and sat at Ludwig's feet.

"Speak!"

"It is reported that a horse and buggy have been found at the entrance to the caverns at Littleton. It has been confessed that Sam Littleton has taken over and found shelter within the caverns. These caves do lead straight through the mountain to Galala Valley beyond!"

"You have found all this out in one run?"

"I went to seven other villages but there had been no word at any of them."

"And what of Village Littleton itself?"

"It has offered its devotion."

"Then Samantha's fate is sealed. We go around the mountains on foot and meet her on the other side in Galala Valley. Spread the word and we will gather an army. No doubt Sam has the powers of the Underworld on her side so we will gather our own force of power. White Power! THIS MEANS WAR!"

"War?"

"What's Galala Valley?"

"Passage through the mountains?"

"I know that you all have a lot of questions."

"You are actually going to start a war?" Tralina asked.

"The only enemy we have is Samantha and her Underworld powers. Whatever she has. Village Littleton has given us their devotion and with that we have the right to go after her."

"I understand that. But I have never heard of going beyond the mountains. It's never been done." Samson explained.

"Generations ago there lived a young female warrior named Gabrielle. She believed that there had to be a passage beyond the mountains. She had dreamed of it. She was also known as a psychic. Her dreams are said to have come true. There is a passage beyond the mountains that leads to Galala Valley. We Elves have taken it upon ourselves to protect the entrance, and the valley, from detection, because of the special inhabitants of Galala Valley."

"Special inhabitants?" Ludwig took a moment to let their curiosity peak.

"The most beautiful and magical beings that ever lived! Centaurs!"

"What's a Centaur?"

"I've heard of them in myths, but surely, they can't be real?"

"That's what you once thought of rust monsters, remember?"

"A rust monster I can understand, but half horse, half human?"

"What?"

"He's right! These beings are real! And just as he describes them. The lower part is the back and hind quarters of a horse. The upper part is the chest and upper body of some kind of humanoid. And there are probably hundreds of them. I know I'm exaggerating. I fact, Galala Valley was named after the only Centaur known to be alive in Gabrielle's time. At least in that particular section of land. Galala, the great Centaur ancestor. Gabrielle wrote whole articles about the Centaurs. They are peaceful creatures, Lawful, who live off the land like any other horse-type. It is said that they are really shape shifters who can take one form or the other of their types. But their reproductive system is special. If a Centaur lives to its hundredth year it becomes pregnant, sort of asexually, and another generation begins. But the mother dies giving birth. So you can understand the need for protection. If a centaur is killed, its generation will end the cycle for that family. And then there wouldn't be any more centaurs on upper earth."

"That's all very interesting. But what does it have to do with us? Are we really going to find this valley and centaurs, or is this all fable? Gabrielle is a fable herself, isn't she? Surely you doubt the ability to find this passage?"

"Not for a second. Gabrielle was a real person. Sure, she's told of in old legends and songs, but all fables have some truth to them."

"So you are actually going to look for this valley and its inhabitants."

"That's right. And you are all coming as escort. Besides, I need all the fighters I can find! Sam's power is stronger than even I can imagine, but we must find a way to rescue Sassy! Or die trying!"

Chapter 2

"Ludicore!" Sam called into the darkness of the passages within the Black Mountains.

"What now?" the dragon beast refuted.

"In here this instant!" The beast inserted itself into the cramped passages, as with much effort he approached. He was forced to tuck his head to prevent banging it against the ceiling. When he reached the ramps entering into Sam's domain he spread his feet to opposing walls and slid down the passage of his choice. Once he stopped sliding the passage leveled out and he stepped easily into the shelter. It was well-lit, and spacious enough for Ludicore to stand straight or sit which Sam gestured for him to do. He sat on his fleshy joints with an impatient look on his face.

"You promised me a war!" he yelled. "All I have seen is peace and quiet. What is the use of Upper Earth if one is not useful?"

"You will be useful in time. Right now all you can do is wait. Sooner, rather than later, Ludwig Wildlife will bring an army of enormous size and many will die. But we will remain unscathed."

"You seem so sure of yourself. I would be careful if I were you. The Beast is likely to turn on you."

"You are useful to me. And I to you. Remember who ressurrected you from the Underworld."

"Probably for your own means. I am powerful. I could kill you with one sweep of my hand. Shatter your skull in one blow. Your brains scattering all over this filthy rock fortress you've built."

"You don't like it?"

"It is adequate. But barely. Who is this female you placed in the chapel? And why is she of any importance?"

"She is Sassy Wildfire, daughter to Ludwig and heir to head sorceress. This is my chance to win my place in society."

"Why not just kill her?"

"And have no reason to war?"

"You could hide her death from them."

"No, I will not be dishonorable. We can get more use from her alive than dead."

"She just lies there unconscious. Surely death is no different."

"You have died before. So you're the expert."

"I do not care to recall my death. It has haunted me since... I don't recall the day I died anymore. Just a very unpleasant feeling."

"Then it wouldn't hurt to die again."

"Is that a threat?"

"Just an observation. I need you too much! If you are defeated I will have no choice but to give in!"

"I will not be defeated. I am a beast of the Underworld. And I am most powerful."

"We are in agreement. What the Elves don't realize and I have always known. Black magic is always stronger than white. Attack or be attacked! That is my motto." Ludicore nodded his head in agreement. "I have your promise. I will obtain the Elven Territories under my control and you will gain Galala Valley. Then you and I will run our jurisdictions with power and will!"

"First thing I do when I get the valley is kill the Centaurs. They are meek and weak and don't deserve to live. Only the strong should survive!" Sam just nodded her head in consent.

"Whatever you do, I do not want you opening the Underworld and releasing spirits beyond the valley. That is yours to do with as you like but the Principality of the Crane and the Territories are mine to conquer. I don't want you outstepping your boundaries."

"We have agreement, then."

"Leave me now." Sam finished. "Go back to your sleep." The dragon just turned and walked out without another word.

"Now tell me about your adventure with Andrew. I understand that Sassy slit his throat." Suchi nodded.

"We took care of the upstairs quickly enough. But the lower level was somewhat of a disaster. Roderick was killed in the conflict but we were able to bring him back."

"Ressurection?"

"How did you k now that?" Tralina wanted to know.

"I've had some success with the same spell."

"I see."

"After all, I'm not just an elf. I am also a cleric."

"Because of your cleric scrolls?"

"My father was a well educated man. He learned Cleric, Elven and common."

"We know that."

"So give me all the details."

"Roderick died of magic missile. We wondered if it really was effective and now we know."

"Magic Missile always works. Any supposition that it wouldn't is false."

"Anyway, we found Andrew's hidden quarters, but he had the upper hand. First thing he tried is paralysis. He almost succeeded too. We were all stopped except Sassy and Lazarus. I saw it all even though I was immobilized. Andrew used invisibility but Sassy blew out all the lights and used her infravision to see him with. He tried to get close to us but Lazarus headed him off and used levitation to protect us. Then Sassy did something that was new to me... actually, I hadn't seen it used in so long that it seemed new."

"Andrew appeared and Sassy began mirror image. As she rotated around Andrew he jabbed the Sassys all around him. But the real Sassy grabbed him from behind and..." Suchi made a slitting motion across his own throat.

"Quick thinker, that girl. Now, where was he hiding, anyway?"

"Outside of Colina to the north-east almost at the Black Mountains. We have a map of the fortress. What amazes us is the mapper's initials are S.W. It's like Sassy herself drew that map."

"Let's see it." Suchi brought out the well-known piece of parchment.

"This is the fortress?"

"You can see all the magic portals are marked and the labyrinth."

"S.W. Serena Wildlife. Head sorceress of Village Wildlife."

"You know the signature?"

"I sure do. Serena was my mother. Lucas was my dad. They lived there when I was young. Barely five years old! My older sister loved to sit in the labyrinth and play hide and seek. There were several ways to go and we ran from each other. This means something important. I believe that Andrew was working with someone on the inside."

"We also thought he had an accomplice."

"Samantha Littleton would be just the person to provide that shelter. It's little known and very private. It's the last place I'd look for Andrew."

"So Sam is his accomplice?"

"I believe so."

"Maybe she captured Sassy to get even."

"At that early stage?" Samson argued. "We had just killed Andrew. How could she have known?"

"No, Sam has a greater plan. So great as to take over the Territories if she can. But she must be stopped."

"Are we enough to stop her?"

"I certainly hope so. But the campaign is already starting to collect. Sorceresses and acolytes both are involved in the campaign and some of them know their own form of black magic. They have learned control. But we may need extra help from the outside. I only hope we can take Samantha alive."

"Why do you care?" Tralina asked. "She kidnapped Sassy. She may already be dead. And you want to spare Samantha?"

"You don't understand. Sam is misled, but she does not deserve to die. Besides, I owe her her own life."

"What?"

"Sam is my step-daughter. Heather Littleton was my first wife. Sam was born before we were married. Heather was weak-minded and some considered her to be disrespectful. But all she needed was a man to take care of her. So I took Heather and her daughter Samantha. Six months later we were married. All she needed was a committed man. She was sensitive by nature and had very strong feelings. She worshipped me. She was so proud to be a headman's wife... even though she remained childless."

"Does Sassy know that they are related?" Ludwig nodded the affirmative.

"For a time Sam adored her step-sister. But her mother's death had hit her hard. The medicine women all came to look at Heather. They all said she didn't have a chance. But Sam sat there at Heather's death bed and watched her mother die. She blamed me in the end. But that's not all."

"There was a conflict between the villages. Amanda, Sassy's mother, was my wife then and the villages were arguing about the position of head sorceress. Littleton believed that Heather should have been head sorceress passing it on to Sam but I had married my second wife, Amanda, as a sealing of friendship between Villages Tralesta and Wildlife."

"Wait a minute!" Tralina yelled. "Did you say Village Tralesta?"

"Yes, my dear. Village Tralesta."

"Your current wife, Sassy's mother, was from Tralesta?" Suchi asked.

"She is the daughter of the head sorceress of Village Tralesta. We made a pact to seal ties. Well, when Village Littleton learned about my second marriage, they were furious. We barely avoided blows. Sam didn't care about rank back then, she was barely an acolyte. But her thirst for attention led her to learn black magic. When I learned what she was doing I refused to allow her to learn in this village. So she went back to Village Littleton and learned from another."

By now Tralina had gone into shock. Her eyes were glazed, and feeling of glee and rage both shook her. She yelled her anger and cried her joy to what she had just learned. Sassy was her cousin and she was in dire straits. She got up and ran out of the hut as fast as she could. Ludwig signaled the others to stay and walked out after her. He found her outside the hut near another fire.

The clouds were gathering, and the sun was hidden, so it had become a fairly cold day. Tralina was shaking with more than cold though.

"I'll kill her! If Sassy's harmed in any way, I'll kill her!" Ludwig sat in front of Tralina and took her roundish chin in his hand.

"I'm sorry. I wish I could have made it easier. There's no way to tell you easily. My wife is your aunt... by blood. And I am your uncle Ludwig."

"Sassy's my cousin. Sassy's my cousin! How are we going to win? Without Sassy we're not the Avengers anymore."

"This has indeed become a family matter." Ludwig responded. "But Sam does not deserve to die."

"If she has hurt Sassy in any way I will kill her with my own two hands."

"We must all pray for a happy ending. But nothing's guaranteed." Suddenly Tralina's arms were around his shoulders and she cried her anguish. She wished she'd never learned the truth. Chapter 3

Ludwig sat with Tralina talking in private for about an hour then left her to her thoughts. Thonolan approached twice before she motioned him to sit.

"There's something I don't understand." he opened. "You are a daughter of a medicine woman. Sassy is the daughter and heir to head sorceress. That means that the medicine woman and sorceress of Village Tralesta were sisters."

"They usually are." Tralina responded. "The headman's daughters all have their choice. My aunt Amanda was older, first born, so she was heir to sorceress. My mother, Rachel, was younger, but she also wanted to serve, so she learned medicine. I was to follow her footsteps, as I am the only child!" Thonolan nodded understandingly.

"This must come as a shock!"

"That's an understatement!"

"I've been talking to Suchi about the matrimonial. I'd still like to go through with it."

"I'm not sure that's such a good idea right now."

"Well, I am sure." Thonolan insisted. "It would be a way to relax. Try to forget for a time."

"I'll never forget, now that I know. Sassy was more than just my teacher. She's my kin. How can I forget that?"

"Maybe you shouldn't. I think that you are meant to know that now."

"But why now? Why weren't we introduced as children? I've always wanted a sister. That is what Sassy is to me. The sister I never had!"

"And I understand that. But our matrimonial would be a sign of hope for the entire territories. Also, I want to be with you in your time of trial."

"You always are."

"You know what I mean."

"Sexual relations. It's bound to happen eventually. So why not now?"

"I love you and I want you. I always have. But matrimony is important, too."

"Marriage should not restrict our feelings for each other!" Thonolan nodded his head.

"I want you. I want to comfort you. Sex would relieve a lot of stress for both of us."

"What's Suchi say?"

"What can he say?"

"I could get pregnant."

"All the better! I've always wanted a family of my own." Thonolan touched lightly on Tralina's face. "I'm dying to make love to you."

"Just to relieve your needs?"

"Yours too. I see the way you look at me. With your staring eyes so blue and hazel. They change with your moods. And there are times when I see desire in those eyes. You are very sex driven. So am I.

There were times coming here that we would stare into each other's eyes and dream of really sleeping in the same bed roll making love all night without pondering eyes."

Tralina leaned against him and put her lips to his. The kiss was long, and the yearning began for both of them. She put her hand in the back of his neck and drew him closer. She could not get enough. It would not be enough until they were both satisfied. Someone nearby cleared his throat. Looking up they saw Suchi standing over both of them. Kneeling down, he wagged a finger toward both of them.

"This is overpowering for both of you. I have decided to forego the stricture. Ludwig has decided to perform the matrimonial himself. If it is still on, that is!"

"You bet it is!" Thonolan confirmed. Tralina remained silent.

"I have arranged for private quarters for both of you. This love of yours is the rare kind, and I have no right to restrict your love life! I'm only sorry you two waited this long. The strength and desire are not worth refusing!"

"And pregnancy?" Tralina asked.

"You will have nine months of grace before labour. This campaign will be over long before that."

"And afterward?"

"I think it's time the Avengers went separate ways. Times are changing. I knew this union would not last. But at least we all profited from our union."

"If Sassy's dead..."

"We have to think positive about this. We can always hope."

"And pray." Tralina suggested.

"To any god who's out there." Thonolan finished.

"Come. It is time!"

"Time for what?" the lovers both asked.

"Time to fulfill your destinies." Suchi waved for them to rise. They followed him to a hut made of heavy white birch twigs. Suchi gestured for them to enter. He followed.

"From now on these are your quarters. Your private quarters! The Avengers are disbanding. Tralina will take over the position of medicine woman of Village Wildlife."

"You're joking!" Suchi just walked out. Tralina's eyes glowed with excitement and the love in her eyes gleamed. Thonolan just kissed her and pushed her down.

"What are we going to do next?"

"Do you trust me?"

"Infinitely." Tralina was wearing leather armour which slipped over the head. She sat up and gestured to remove it.

"Let me." Thonolan grabbed the leather by the shoulder and pulled it over her head. Thonolan was wearing a fitted robe with many pockets. She pulled it over him. His manhood stood in open air and it was swelling. The shaft was six inches long at erection and she opened her eyes in amazement. Thonolan took a long glance at the body of his fiance. She was flat chested with the breasts small and virgin. They both lay back, Thonolan on top, his hips to the side.

"Kiss me!" she insisted. He responded with a kiss on her bottom lip and she pulled him to her once more. He nibbled her lip and lay his hands on her hips, drawing them up to hold her gentle mounds.

Suddenly something jerked him to awareness. His loins were screaming for release. He nibbled her aereolas and suckled, moving his hands to the sensitive spots underneath. Making tiny circles, he brought her to awareness.

The grinding feeling was strong in both of them and Thonolan was out of control. He wanted her like he had never wanted anyone. Both of them were essentially virgin, and both marveled in the feeling true love gave them.

"I think I'm ready." Thonolan grunted. Tralina laughed.

"If you wanted to you could have taken me right away."

"Are you ready?" he asked.

"I've been ready since I met you. But now..." She just nodded her reply. Thonolan rose, and laid on her, placing the manhood where it belonged. Love and trust lay in her eyes despite the fear of pain. He lay his hands on her and rubber her to wetness. Then he thrust slowly and she opened. But a small shout of pain still revealed itself. He dug deeper and deeper and her pain was obvious in her eyes. But she did not shout again. He was deep within her now, with his shaft deep and full. He pulled and pushed with sensitivity and care. She could feel him within her, and she marvelled at the feeling. He twisted and moved bringing exotic feelings of satisfaction and happiness. Then he delved to the end of his shaft and she shouted with his depth. She felt one strong surge and then wetness. But he was not done. He kissed her stomach and his tongue made circles. He went lower and lower and just the anticipation brought her to climax. He delved again and increased the motion. He felt wetness once more and it came. She could feel his wetness enter her on its way to life. Only would she conceive?

They both lay there, spent, longing for it once more. But at the same time too tired to continue. They just lay there and enjoy the closeness. Thonolan moved and lay beside her.

"That was amazing! I always thought it would be good, but never this way!"

"Our love makes it twice the feeling!"

"You pregnant yet?" She just shrugged.

"We'll see."

"I love you." She kissed him, knowing no matter how many times they did it, they would never have enough.

"I'm going to make you a promise. I'll never refuse you. I couldn't even if I wanted to. And I don't."

"You're too much!" he responded.

"I love it when you're on me. It's like we are one being when we are like that."

"That's what sex is all about. A union. A blessed feeling of bliss which never ends."

"Are you going on the campaign?"|

"Aren't you?"

"I just want you near me."

"That will always be. I promise you!"

Chapter 4

Morning came with the blessing of a clear sky as the sun edged over the horizon. Tralina kissed Thonolan's smiling face and rose. She dressed and exited for a morning swim in a nearby lake. After she washed thoroughly, she returned to the village and watched as people came and went while preparing the morning meal.

"Tralina." Ludwig saw her and approached, taking her hand in his. "The matrimonial will be performed before the morning meal. It will be a special ceremony conducted in Elven custom. It will be extravagant as pertains to your lineage."

"And Thonolan?"

"As your suitor, he has the same rights and privileges of any elf. I have already discussed this with the council. The Elders arrived last night, so the council will be in conference most of the day. I will plea for Sam's life. But the decision will be made in committee."

"The campaign will start soon?"

"Everyone is ready, and some have already arrived. Thirty-five sorceresses with acolytes. Fifteen headsmen and ten swordsmen. All mounted. The passage is some distance away and horses would get us there faster."

"What about the Avengers?"

"The troup is coming with us and Eugene has pulled in his resources as well. Fifteen fighters and clerics, as well as ten magic users, thanks to Lazarus' connections. The trip starts at sun-up tomorrow."

"Are you sure I should come?"

"It would be best. Yours is the first face I want Sassy to see."

"Then it's decided." Tralina walked to her quarters and ducked entry. Ludwig moved on to make arrangements for the following day.

They assembled in a group which consisted of 35 headmen and elders. Sitting beside a raging fire in the prairies established for meditation and practice. Ludwig rang the ceremonial gong for silence.

"We are here to decide the fate of Samantha Littleton in her treachery of kidnapping our beloved fellow maiden Sasselia Wildlife. I will now open discussion.

"Sam has a lot to pay for." Garnen of Village Lakely spoke. "She has purposely broken the treaty of companions and kidnapped one of her own. She must pay with her death."

"I don't see it the same way." Ludwig informed. "She is misled and confused. It's all my fault."

"Don't be ridiculous." Garnen replied. "She had her chance, and she made her own decision."

"I still don't believe she deserves to die."

"She has deliberately ignored the treaty of companions and kidnapped her own adoptive sister. She must die!" Garnen's voice rose in pitch and he yelled back at Ludwig. "She must die!"

"All this because of the slight against your son, Garnen. Because Sam refused your Erik's hand in marriage you would sentence her to death." Ludwig was raising his voice in anger. "Don't take your anger out on her. The issue here is the campaign and Sam's fate." Ludwig finally calmed himself.

"We must use discretion." the elder named Perkodes responded. "Ludwig is right in noting Garnen's prejudice. But the situation of Ludwig's position must also be an issue here. Sam is Ludwig's first wife's daughter. For her sake you might want to spare Sam's life."

"I understand that and acknowledge it."

"A life for a life." Perkodes responded. "We will wait and decide Sam's fate after she is captured. A life for a life! If Sassy is dead we will certainly kill her. But if Sassy still breathes Sam will be spared. If the underworld does not kill her first!"

"On to the next topic. The marriage of Tralina and Thonolan."

"The villages have never seen an iter-racial marriage before. Will it really be legal?"

"It's preposterous. Two races cannot procreate."

"I tend to differ." Ludwig put in. "We're all human. It's not like a combination of human and dwarf. Besides, dwarves tend to lead a life of custom. They're so ingrained that the thought of another race is beyond their belief."

"There are differences."

"Short stature and pointed ears make a difference? All females bleed! Why should the difference be so great? After all, we are all human."

"Humanoid, you mean."

"Royalty has the right of choice."

"And their offspring?"

"Will be human. After all, pointed ears are only a possibility. In the villages they are dominant but with Thonolan, who knows?"

"It might have pointed ears and Thonolan's height."

"Or Tralina's shorter stature and human ears."

"The question is the future of such an offspring. Will he/she be criticized? Will prejudice have a part in its life?"

"The mutants in Colina can't be much worse. And they are all proof of the fertility factors."

"I will not tolerate prejudice in my village. This is a family unit, and I can't believe that anyone would be prejudiced against any offspring."

"Children are children, no matter how they turn out."

"And Sam?"

"What?"

"Sam is one of our children, too! She also deserves to be accepted."

"Not after what she's done, surely!"

"She's deluded and confused. She deserves to be given a chance."

"A chance for what?"

"To redeem herself. That is... if Sassy's still with us. But if she's not, Sam will surely die."

"That's the deal."

"Sam's not cold hearted. I don't believe she would commit murder. Least of all to kill her own sister."

"You're convinced that Sam will be spared."

"I certainly hope so."

Charles Littleton sneaked up to the house of Sandra Littleton, wife to Chuck, and peaked in the window. Sandra was speaking to Samantha in a calm voice. A third voice was suddenly heard from another room, a young-sounding, male voice.

"Now you've done it." Sandra said. "You've awakened Joseph."

"I'm taking him with me, anyway." Sam responded.

"Into those dark caverns? Are you sure? Can you take care of a child and fight at the same time?"

"He's two years old now. He's off the baby food and finished nursing. He's even potty trained. He'll be okay."

"He's your son of course, but I recommend against it."

"You've just grown attached to him."

"Maybe and maybe not." There was a sudden shout of crying from the next room. Sam walked past Sandra and entered the sleeping area. Charles moved to the bedroom window. Picking up Joseph, she cuddled him in her arms. The youngster clung to her and smiled with glee. Charles got a good look at Joseph. He looked as normal as any baby. Then he noticed the human ears. His ears were not pointed.

"I'll see you later." Sam finished as she exited the living quarters with the boy in her arms. Charles ducked down and ran from behind the house in the direction of Village Wildlife.

"We are here to conduct the matrimonial of Tralina of Village Tralesta and Thonolan the Brave." Ludwig stood facing the fiances as the whole village looked on. "We will begin with the ceremonial firing." Amanda Wildlife approached and threw a fireball into a wet fire pit. The pit was damp, and still, the fireball was strong enough to dry the tinder and make fire start.

"Let the purifying fire of magic bless this holy union!" Amanda declared. "By the blessing of Village Wildlife I hereby declare the joining of these two people into union. Anyone who stands in their way may speak now or never. Does the whole village agree to this joining?" Everyone said a loud affirmative. "Village Elders of the Elven Territories, do you agree to this union?"

"We agree to the union and all the chaos or pleasure it may bring. The joining is an unusual one but we will give all our resources and support to protect this union and new-found family." Perkodes, as Head Elder, spoke for the whole group.

"Then let it be done." Amanda backed away and Ludwig returned to his place.

"Tralina, do you solemnly swear to share only with Thonolan and to provide for his every whim?"

"I do."

"Thonolan, do you swear to uphold your station as provider and agree to succor your maiden's life?"

"I do."

"Then, as headman of Village Wildlife, I declare you man and wife. Walk around the ceremonial fire three times, and it will be done!"

As the man and his woman walked around the fire the first time Amanda put sulfur into the fire in small amounts causing smoke to rise and penetrate the very sight of the mates as they walked.

"By smoke of sulfur I consecrate this marriage and any future it may hold." Amanda vowed.

As they went around the second time, Rachel Tralesta, Tralina's mother, came with rocks and threw them into the fire.

"May you stand on solid ground made of a mountain of rocks and may never a flush of water move you." Tralina and Thonolan completed the second circle as Rachel stepped up to hug her daughter. "I have taught you all I know of medicine and you will learn much more as you begin your new life as medicine woman for Village Wildlife." Ludwig gestured to them to complete the third circle. This went uneventfully. Then Rachel walked up and took Thonolan in hand, showing him the platinum dagger he had been offered before.

"This dagger belonged to Lazaman, my husband, who is now dead. You are to give it to the son which your wife now carries."

"How can you be sure?" Thonolan asked. Tralina was awash with disbelief. Ludwig took over.

"As you have completed the cycle of cleansing, I hereby declare you man and wife."

Thonolan and Tralina kissed and the ceremony ended with cheers of good luck.

Someone was now rushing through the crowd at a quick pace. He stopped near Ludwig and knelt at his feet.

"My lord, I must speak." The crowd was loud, so Ludwig could not hear very well.

"Silence!" he called. "This man will speak."

"I am Charles Littleton and I have important information from my village."

"Then speak."

"The maiden Samantha Littleton must not die! She has a son!"

"What?"

"The child is two years of age and named Joseph." Immediately Perkodes walked up and spoke to Ludwig.

"This means more than just death. This means her salvation!"

"Two years old? Who's the father?"

"I got a look at the child and the first thing I noticed was human ears."

"A hybrid." Perkodes concluded.

"Who would be her lover?" Thonolan asked.

"That is obvious. The only person close to her had been Andrew."

"Andrew's the father. That's what I conclude, too." Ludwig admitted.

"So how does this affect us?"

"A child has lost his father. We must all make up for that. Now we have no choice but to spare the mother."

"She is a mother. That means we must do all we can to spare her life." Ludwig continued.

"How could she join up with a Chaotic?"

"Loneliness drives one to desperate measures. She feels herself an outcast. What better than to bed another outcast?"

"We killed her lover. How does that affect the Territories?"

"It means that not only must we try to protect her, but we are indebted to nurture her and her son. Support them both as long as they live."

"I agree." Perkodes confirmed. "We have no choice than to accept her back into the fold."

"She's Chaotic, isn't she?" Tralina asked.

"No, she's neutral. The whole village vowed neutrality at the proper age. She does not want to be mean... she's just horribly lonely and feels betrayed. Just like Heather. Why do the trials of our forefathers fall on the children?"

"Now she's just as lonely as before. Except for Joseph."

"That's why she took him, I bet." Charles stated. "He has been under the charge of Sandra Littleton who had agreed to raise him in the first place."

"How did the child take to the reunion?"

"He was all smiles! He recognized her immediately!"

"They had not lost contact. They have always known each other."

"That's possible but it was kept quiet." Charles concluded.

"Is there any birth record?"

"Only in Littleton and that's under tight seal."

"Then we have no choice. We have created a burden for ourselves and we must concede. Let's all hope for the survival of the both of them. I hope she does not have to pay for her menace with her own life."

Chapter 5

" Mommy, why do you have pointed ears?"

"Because I was born that way!" Samantha answered.

"Everyone has pointed ears but me."

"That's not true. Your dad didn't."

"He sounds like a good man."

"Actually, he wasn't, I'm sad to say. He was chaotic."

"What's chaotic?"

"Come here." Samantha gestured and Joseph sat willingly on her knee.

"Let's see. Chaotic means evil, neutral means you're willing to bend the law and Lawful characters are just gullible. Anyone can swindle them. They obey the law without question."

"What are we?"

"We are both neutral."

"Why was dad chaotic?"

"It was his choice."

"Is he dead?"

"He is now. It's not that long ago that he died, and chaotics tend to get killed. They're not liked very much."

"His name was Andrew, wasn't it?" Samantha just nodded her head. "Why was I born?" Joseph surprised her with this question.

"Do you know what loneliness is?"

"I do! I'm lonely for you a lot. I miss you when you are not around."

"Why's that?"

"Because you are my mother. And I love you!"

"True. And I love you too."

"Why was I born?" Joseph wanted an answer.

"Because I was lonely, I guess."

"Why were you lonely?"

"Being an outsider does that to you!"

"Why are you an outsider? You're just like everyone else."

"We're a couple of outsiders, alright! That's what happens when you're not born in wedlock."

"We're bastards." Joseph replied.

"You have way too much of your father in you! But that is the correct term."

"Why do you stay here?"

"For protection. Besides, I always wanted my own place. It's not too bad." Joseph shrugged his shoulders.

"It's livable."

"Yeah, it is."

"Why's that statue standing so close to the entrance?"

"That is not a statue! That is a real dragon!"

"Then why doesn't he move? Dragons breath fire and sulfur."

"He's resting. He does that a lot! There's nothing else to do in this dumpy place!"

"I thought you liked it!"

"I do. But sometimes you need sunshine too."

"What's the dragon's name?"

"Ludicore!" Sam replied honestly.

"Is he our guardian angel?"

"Hardly! He's chaotic. I wouldn't cross him if I were you."

"Why do you deal with chaotics?"

"I'm afraid that is a failure of mine. It is because darkness is stronger than light."

"Is it really? I wonder."

"Come. I will introduce you to Ludicore." They both went outside and approached the beast.

"Ludicore you old beast. Wake up!" Ludicore awoke as his stony body turned to flesh.

"What do you want? Who disturbs my sleep?"

"I want to introduce you to Joseph Littleton, my son." Ludicore eyed Joseph up and down and then just turned away.

"You are rude!" Joseph opened.

"I'm also ugly!" Ludicore replied.

"That's an understatement!" Ludicore just laughed.

"I like you kid! But don't take me for granted. You never know how this war will turn out."

"War?" Joseph was perplexed. Sam was visibly annoyed at the dragon's implications.

"Go wait for me inside." Sam asked.

"Yes mother." Joseph left and Sam immediately turned on the beast.

"Don't mention the war to an innocent. He's my son, and I love him."

"Just like you loved Andrew?"

"Andrew was a fool. He didn't know when to say no."

"And you used him as a pawn to sooth your own loneliness."

"Joseph is my happiness. I don't want you making empty threats."

"I assure you, they are not empty!" Sam was struck dumb, so she just passed him a glare and walked away.

"Mommy, why are we at war?" Sam just walked from the entrance and sat down. She gestured once more for him to sit close.

"We are at war because I have had enough of being denied. My father refused me my training so I went and learned in my mom's home village. Littleton."

"She was my grandmother?"

"She was. But she didn't live long enough to know you. Her name was Heather."

"So you're just fighting for your rights."

"I'm fighting for my life! Ludicore is part of our protection but he cannot be trusted. When the time comes he will turn on us. Of that I have no doubt."

"So what are you gambling with? What do you have that your dad wants?"

"You are too smart for age! I have insurance. Come into the chapel with me."

"Chapel?"

"Come." Sam walked into the middle of the room and walked through a solid wall.

"Mom?" Charles was perplexed once more. Sam leaned half way through the wall and waved. Joseph joined her.

Beyond the wall was a slightly spacious room which Sam used as a chapel. The room was decorated with torches which were changed frequently. In the middle of the room lay Sassy Wildfire aka Sasselia Wildlife. She was raised on a dais and in deep sleep.

"Sleep well and long darling!" Sam quoted as she gently stroked Sassy's cheek. "It won't be long and you will be free again. I promise!"

"Who is she?" Joseph asked.

"Your aunt Sassy. Sasselia Wildlife. Daughter of Ludwig Wildlife your grandfather."

"She sleeps so soundly. I can hardly hear her breathe."

"She's under a sleep spell. I forced her into sleep."

"What's going to happen to her?"

"She's just my insurance. When I get what I want I will release her."

"As long as this is not permanent." Joseph replied. "I'd like to get to know her."

"I'm sure you would!"

"What's to prevent Ludicore from turning on us? The thought of this war frightens me."

"It's more like a campaign, really. No one will get seriously hurt... at least I hope not."

"Ludicore's the outside element in this mess. I don't think we can trust him."

"I know we can't. But it is a gamble... the biggest gamble I have taken in my entire life. But I deserve to be head sorceress. Sassy is the heir to that position but it really have been me."

"So you're willing to gamble your life and mine as well!" Joseph replied. "There are other ways to get what you want."

"This is the only way! No one will listen to me otherwise. I'm an outsider remember?"

"So you kidnapped your own sister for insurance. I can forgive you for that but as for Ludicore... I can't see the end of it."

"Neither can I."

"Take good care of Aunt Sassy. That is what I want. She's better as an ally than as a prisoner." Sam just ignored his sarcasm.

"I have something else I want to show you." Sam led Joseph out of the chapel and to the far wall where there was a flight of stairs going down. He followed her to the lower floor.

"We came through here on our way."

"And I have a surprise for you!" Sam walked to the right a distance and pointed to the cages at the far end. "These are my pets." Sam cast a spell which allowed the pets to speak.

"Hello, madam. Is it meal time already?"

"Yes, Saliva. It is meal time!" Saliva was a female lion and her mate was named Rodren. Three black panthers sat in another cage. They remained unnamed but had all the comforts just the same. After Sam had handed out portions for her pets she pulled a lever and the animals were released into a large corral. Saliva came out for an affectionate caress.

"Sweetheart, I'm going to need your help." Sam stated. "You are the leader, so I speak to you."

"How can I serve you?"

"Can you take out a dragon?"

"I can try. Maybe go for the hind legs."

"If you can get close enough that is."

"Don't worry. We'll try our best! I take it you are scared of that monster? Well, we will gladly try to take him out. I hear dragon flesh is sooo good."

"Actually, he's rotten."

"To the core I bet." Sam laughed.

"You are such a comfort! I don't know what I would have done without your company these past two years."

"Don't forget my pals! They like to be stroked too."

"How could you forget?" Joseph sat down and leaned on his mother. Soon he fell asleep.

"You better take the tike upstairs and put him to bed. He's pooped."

"You're such a comfort, Saliva! I hope everything turns out right."

"So do I." Saliva closed. Sam got up and carried her son to bed.

He awoke and looked in. Everything was intensely dark in there. Sam had darkened the cavern for the night. Ludicore had been thinking all day and now was the time to put his thoughts into action. If he was to die he would take someone with him. He slid down the passage to a stop. Stooping, he entered the chapel. Ten minutes later he came back out and went back to his post. He had feared his comings and goings might have been detected... but it was most fortunate. His first step in this war had finally been accomplished. Step two was Samantha and her offspring. He would not be the only one going to the Underworld.

Chapter 6

"Everyone's ready."

"Mount up."

Suchi mounted Sunshine and the campaign began. The trail through the forest was a small one, wide enough for two riders astride, but that was all. The troupe took up positions in twos. The Avengers rode as one group in their own chosen position. Thonolan rode with Tralina behind him among the others. The attire was leather armour where appropriate and the magicians wore elegant decorated robes.

Lazarus rode with his fellows from his home guild. Ludwig lead in the front and Amanda rode directly behind. The troupe started slow and picked up the pace as people chose their places.

The group travelled through the villages to the north. At each village there were mothers, tikes and teens who were being left behind at a moderate pace. They were off to find the famous passage to Galala Valley. The troupes were cheered at each village.

Nights would be spent at one of the villages until they left the forest. They traveled for three days until they came to the second last village in the Territories. This was where they would stay that night.

Tralina was waking up sick to her stomach most days and as time passed her breasts became very sensitive. She was indeed pregnant! The couple started smiling with knowledge as they rode. Nights for the couple were spent in private quarters which were gladly provided.

Soon now they would meet the valley. Locals told that the valley lay two days' travel beyond the forest. The area beyond the forest was lovingly called "The Gate". It was a distance to ride and it was considered implausible to walk the distance. Many considered The Gate to be too sacred to desicrate.

This night they stayed in Village Durken, the second last village before the forest gave way. Everyone dismounted and unbridled their horses leaving them to graze.

The village had arranged for a massive meal in honour of the campaign. Tralina could not eat much of anything. She was constantly accompanied by two fellow medicine women. The baby seemed to be holding his own and there was no sign of bleeding. It seemed that the baby would indeed be healthy.

"Remember, it will be a boy." Rachel told Tralina as they ate. "He will be a healthy child, too!"

"I hope so. But the thought of child birth scares me! Women with complications have died."

"That is only in worst case scenarios. But yours seems to show the go-ahead.

"I'm still sick sometimes. Other times just the thought of food makes me sick!"

"That happens at times. But it's normal. There's no sign of bleeding and no sign of miscarriage. You are healthy and strong. Everything is and will remain fine. Thonolan is a strong father."

"I'm holding my stomach again!" Tralina said as she looked down. "How come you never told me about my connection with Sassy? I didn't know Amanda was the sister who left before I was born!"

"I had not known myself! After all, Village Wildlife is where it is and we are so much farther to the north. I had heard the name of her new home, but as far as I was concerned, it was another world away. The Principality was a name which was seldom referred to, and even then, only by travelers. I never expected to meet my sister again, and we certainly could not keep in touch."

"Dad told me he loved you! Before he died!"

"That old fool could have lived longer if he hadn't put it in him mind to leave! But you knew enough to take care of him until his demise. He knew he was dying. That's why he took you on that trip! He thought an outsider could teach you better than anyone in the Territories. He was obstinate about that. He wasn't coming back. And he hoped to find you a mate. God bless his soul, it seems he succeeded." Tralina nodded her head.

"He taught me my first spell. Fireball!"

"That old man never knew much but he picked up part of the knowledge. He was a strange one, learning magic and being male. Anyway, he is in the Underworld now. Or maybe in heaven waiting for me!" Rachel put aside an extra piece of wood after stoking the fire.

"Do you think you would ever remarry?" Tralina asked. "Dad was old when I was born and he was really old at his demise. I'm sure someone would take you!"

"I've been thinking about that and I've decided to give it a chance. Maybe with time I will meet someone."

"That would be a blessing!"

"Aren't you asleep yet?" Suchi asked of Samson. "That wine usually knocks you out after the first bag. This your second?"

"Still my first!" Samson responded.

"Then I see. No rush, eh?"

"Sit." Samson asked. "I'm concerned about my own future. The group is separating, and I find myself at a loss for what to do with my future."

"Understandable."

"I think it's time I found myself a wife! With my dagger I can claim my title at any time. The war seems to be dying down I hear! So after we get back I will go to the Dwarven Terrs. I think it is time. Well beyond time in fact. I want to thank you though. I never knew how lonely I was until I joined up. But now I think it is time I went back. I'm an old man and am still running."

"Understood. You deserve a good mate. And offspring too! You could take one of your mates!"

"What?"

"Sure, an Elven wife would be fine."

"Don't be ridiculous. I will choose my own wife thank you! And one of my own too. You know how we feel about mates. We are strictly inbred. But maybe that is a bad idea, after all!"

"Big brother?" Roderick gestured to Roger. "I think it's time we talked about the future! A lot has changed. The troupe is breaking up, and I don't know why."

"I think Sassy's problems have shaken our bosses. They're afraid of messing up again!" Roger figured. "Understandably."

"What will we do after this? Go back to the guild?" Roger shrugged his shoulders.

"I think the guild will not profit us anymore."

"Then what?" Roger got up and approached Suchi who was still talking to Samson.

"Can we talk?" Roger asked. Suchi rose and walked to sit in the Bloodbaths' camp.

"You're concerned about the prelonged future." Suchi guessed.

"What will we do with ourselves after the campaign?" Roderick asked. We do not want the band to break up."

"We don't have to completely. I'm just concerned about the weaker people! Just now I realize how weak Sassy is in reality. Unprepared, she could be killed... or kidnapped."

"What have you got planned?"

"Lord Brunweger and I have been talking... about all concerned. You two will stay on in Colina as gate guards! The protection of the town will be in your hands. Plus he is planning to man a force of fighters who will keep the peace. With violence the last alternative. Also, he wants to band chaotic thieves from within city limits. They will have to be taken out. That's Roderick's immediate consideration. How to move them out and to decide whether or not to use force. And support a force strong enough to obliterate if necessary."

"War!" Roderick exclaimed. "Those chaotics belong in the Underworld, alright! And I'll put them there."

"Samson is going back to the Dwarven Terrs. His home. The war is over, and he is ready to return to start over!"

"Aaalright!" Roderick yelled as both fighters gave fivers with their hands. "He's going to be King!"

"I'm worried about Sassy. Will she survive?" Suchi confessed. "I have had a request for her hand in marriage. Alvin Guard is a sailor in the Elven Guard crew working the river and ocean front. He has traveled the waters widely and he is ready to settle, he says. He is planning to settle in Village Wildlife."

"Who is this man?" Roger asked. "He does not know her yet. How can he even ask?"

"For your information Alvin and Sassy were play mates. Alvin is from Village Littleton, Chuck Littleton's son. He has always kept tabs on her and he has asked for her hand in marriage."

"Boy, Sassy's going to get a wide awakening when she recovers. She's engaged."

"Does Ludwig know?"

"He has known this man most of his life and he does approve. In fact, he's offering to make Alvin headman in his own place."

"Sassy will be Sorceress?'

"In place of Samantha until she learns what she needs. Then Sam will be head sorceress."

"Then why the war? Why not give in?"

"Sam has always abused magic. She has to learn that you cannot deal with black magic without knowing how to protect yourself."

"From backlash!"

"It's not a matter of power so much as a matter of protection. Defense. Sam has no defensive skills to prevent backlash."

"So she has to train and learn defense!" Suchi nodded his head.

"Then she will be able to be head sorceress after all." he explained. "It's that simple."

"I guess we all have to learn the hard way."

"But at the stake of three lives or more? Sassy, Sam and her son Joseph. They could all die. Our job is to protect all three at all costs. We don't know what exactly we're dealing with yet but time will tell. It can be very dangerous. In fact, this whole group of warriors may not be enough to end this well. We are only grasping at straws right now."

"The heavens have mercy on us!" Roderick quoted. "Or we may all go to The Underworld!"

"Exactly! Lazarus is going to teach at his guild. Samson's going home. You two will have your hands full. Thonolan and Tralina will be staying in the Terrs. And Sassy's staying as well."

"What will you do with yourself?" Roger asked.

"I will be in charge of Colina. Eugene's making me mayor in his stead. He will be raised to the rank of Governor. He will be in charge of the principality as a whole. He will be secretary to Leo Anasthasius himself! The prince."

"He will move to Anasthasia then?"

"That's where he will be. And I will be mayor. Of course that's in the near future but not too near. We can at least finish this campaign as a last mission."

Chapter 7

The campaign began again full stream ahead the next morning. They rode til midday before they reached the last village. Passing through, they did not stop. Toward dusk they reached the end of the forest beyond which spread vast prairies. To the west could be seen the peaks of the Black Mountains. To the north stood vast, hilly prairies and the Calico Mountains could be seen some distance to the far north. These were said to join the Black Mountains, but that was in fact a falsehood. When everyone was out of the dense foliage Ludwig called a halt.

"Where to now?" Suchi asked.

"We follow the Black Mountains to the pass which is admired as The Gate. Then we decide where to go from there." Ludwig responded. "The mountains also may be a good source of fresh water."

"How far are the Black Mountains from here?"

"I'd say half-day travel at top speed. But we don't need to rush, so we'll take it kind of easy."

"Shall we continue or make camp here?" Ludwig turned to the lowering sun.

"We go a distance yet. At least get away from the forest. We still have a half-hour's light."

"Move out!" Ludwig yelled as he moved his hand across the horizon. He spurred his horse to a run as the others matched his pace.

The campaign trotted at their strongest pace for the half-hour, turning toward the chosen line of mountains. At the end of that time the sun was gone beyond the horizon which darkened the landscape and prevented further travel. Ludwig called for a halt.

"This is where we camp for tonight. Dismount and unbridle." Everyone got to work setting up camp. The women soon started a meal and the men discussed strategy.

"The Gate is to the north of the Black Mountains. If we follow them they will lead us right there."

"Then we keep on going in the same direction until we reach them."

"Then we follow the mountains to the north."

"Two days' travel should lead us to the passage."

"Then what?"

"We take our chances. We know that the caves at Littleton come out at the opposite side of the Black Mountains. Based on the location of Littleton in the Terrs, we should hit the cave in..." Ludwig did some mental calculating. "ten days, maybe twelve. And that depends on passable landscape. We will or may have to cross Galala River at some point somehow, so that may slow us down, plus any detours we have to make. Remember, men, Galala Valley is foreign territory, so we must remain aware of our surroundings."

Someone yelled the approach of a troupe of horses. Ludwig turned to see six Elven warriors dismount. He looked through the fire to see a young man approaching. His hair was blonde, brownish in fire light, and his ears were indeed Elven. He adorned blue eyes and a pudgy nose similar to Ludwig's own. This man was known only as Alvin Guard.

"Alvin Littleton reporting for duty, as with your permission, I'd like to come along."

"You are welcome, my friend! We will need every man. This is my partner, Suchi Evil Killer. He is a cleric."

"I've heard good things about you, sir!" Suchi shook his hand. "You are Sassy's betrothed of course."

"She doesn't even know about the arrangement. I think I should propose before we discuss details."

"I heartily agree. Come, sit by the fire." Suchi called his mates over and everyone sat. Suchi introduced his troups to Alvin and they discussed the situation.

"You're Sassy's betrothed?" Tralina asked.

"If she would take me that is."

"Suchi said you are from Littleton."

"I'm Chuck and Sandra's son."

"Have you heard the news, Al?"

"What's that?"

"Your parents have been taking care of a two-year-old. Samantha Littleton's son."

"Sam's the one holding Sassy captive."

"I just came from home and they were both very quiet. The atmosphere was even stale. I've never seen them so lost for words."

"That's probably the way it will stay, too. With Sam it is a very touchy case." Ludwig explained. She has called on The Underworld for power."

"And she does not know how to use it properly."

"How did you know?"

"I've known Sam Littleton for a long time. We were neighbors for a time. She always was confused."

"Then you understand our situation!"

"I do. Sam is obsessed with power when she should be concerned about control."

"We are at a loss as to what to do. All we know is that we have no details, and we must try to help instead of attack. Sam must not die!"

"Sam picked up her son some time ago." Tralina mentioned. "He is also in danger."

"You have small details... gossip... but no important details. Facts."

"We must play this by ear."

"I agree. We have no facts so we use caution."

"The moon is full!" Tralina mentioned to Rachel later in the evening. "This is my due date for my period."

"It won't come! You are pregnant."

"I know. I feel different already. It's been less than a month and already I feel the change."

"It's a nine month pregnancy so you have a long way to go."

"What if something goes wrong?"

"There will be signs, bleeding or prolonged morning sickness. Sometimes both can happen."

"I need to keep this baby."

"You will. I still think it will be a boy."

"How can you know?"

"Besides being a medicine woman, I am also an astrologer. I can tell by reading the sky. The stars!"

"That's how you knew it would be a boy?"

"The moon is out tonight and the phases of the moon determine the sex. Depending when the moon is in each phase it can determine sex, personality and health."

"You read the stars." Rachel nodded her head. Tralina lay down from her sitting position and became quiet... watching the sky, but unaware of her position, she soon fell asleep. The troups all slept among the stars with just ground covers for the damp grass.

The next morning they waited until the sun was high above the horizon before they saddled up. Ludwig figured they would go as far as the Black Mountains that day.

They rode til midday and then had lunch. During that stay they saw a herd of wild horses. It appeared that horses ran rampant in these prairies. When they started off again some of the herd began to follow.

The stallion of the herd was a white horse with black stripes. He stomped the ground in challenge to the troups' horses. The horses became agitated and confused. The stallion raced toward the troups but Ludwig cut him off. The stallion, seeing the horses could not be enticed, raced off to his herd to the north to protect his followers. The troups continued on without incident.

That night they arrived at the huge sides of hard, stony mountains. Mountain goats could be seen climbing the steep rock with quiet ease. One took a leap from rock to rock and disappeared over a hill side. Ludwig called a halt.

"Well, looks like we got rid of that herd."

"That stallion was a real beauty. Wish we could tame him!"

"Breaking a horse is an art form. There aren't many that know how." Ludwig responded.

"That mountain goat sure knows how to travel."

"Wish we could travel over a mountain. It might save us some time."

"I wouldn't dare. Horses have tender footsies." There was some laughter.

"What now?" There were cheers of agreement.

"What now?" everyone called out to Ludwig.

"We are at the mountain range so we head north along the range until we reach The Gate."

"How long before dark?" Suchi asked.

"We have about two hours of daylight left. Let's keep going. We'll camp at dusk.

As the troups marched on they saw several more herds of mountain goats and some scraggling horses. Ludwig noted that these were probably loners. Those that were not in a herd or were beaten out of one. There was no way to know what sex the scragglers were or if they were forming their own herds. Not yet.

Ludwig called a halt at dusk and camp was set up for the night. He estimated that they may reach The Gate the following night or the one after.

The truth was that The Gate was still three days away, but on the third day they saw the mountains give way to flatlands expanding to the east, and the Calicos were nowhere to be seen. To the north stood blank flatlands of sweet grass which the horses learned to adore. They had reached The Gate.

"We are here!" Ludwig yelled raising his arms to the sky. "This is The Gate."

"Gabrielle was here." Suchi yelled back. "It's really here."

"Fable proven true." Ludwig exasperated.

He called for the troups to continue. He headed around the mountains and followed them to the west. This was the direction Sam lived in. He took the time to map their passage. He would personally map the entirety of Galala Valley if he got the chance. There were old maps of the Valley, but most were considered inaccurate or undependable and outdated. He would map the Valley and compare his results with those of Gabrielle.

Ludwig called a halt on the second day of scaling the mountains and looked to the north and west. Sweet grass was prosperous as well as wild cabbage, carrots and other vegetables and herbs. The troups lived off the land most of the time these days and only used fats and meats for energy. Ludwig had not known the grass lands would be so fertile.

Looking on at the expanse of grassland, never-ending seemly, it was a world of horses, onagers and goats, and he knew, among the vegetarians was a Centaur or two. Gabrielle had explained their bashful nature and preference to stay hidden until and unless needed. But once they were needed they would appear suddenly and in a hidden form. Perhaps the troups were being watched right now. But not a single horse-form was in sight. Centaurs were lawful creatures who were there when times got rough for the survival of the valley. And right now times were getting there.

"What are you checking for?" Suchi asked.

"This is fairy land. I'm checking to see if we are being watched."

"By Centaurs?"

"Or unicorns or horses. There are no herding animals in sight. I guess that means we are still safe to continue."

"How do you mean? What's a Unicorn?"

"When man is in danger in this valley the Centaurs are always there. If we get in dire straits one will show up. It's a mutual relationship we Elves have with them. We protect The Gate and they respond equally. If we get in trouble, and we will, the Centaur will show. It's just a matter of time.

"So you're watching for him."

"Or her." Suchi nodded his head. Ludwig called to continue with a wave.

Chapter 8

Samantha awoke to bright sunlight streaming in the distance. She got up and decided to visit the chapel. Kneeling, she crossed her legs to listen to the sound of Sassy's sleeping form. As long as Sassy breathed she knew nothing was wrong. But as she meditated, listening for the intake and release of breath, she noticed an abnormality. Curious, she rose and put a finger to Sassy's nose. There was no feeling of exhale. Also, she noticed in the torch light that Sassy's skin pigment was kind of bluish. Sam decided to check for a pulse... there was none. She stood there testing for two whole minutes and only felt two certain beats. Was Sassy's heart tremendously slow at one beat per... minute? She waited some more and realized that Sassy was very close to death. She had, in fact, entered into a coma.

A chill went through Sam as she wondered how this had happened. Her own sleep spell had been first level, the only sleep spell she was in possession of. This should not have happened, and she knew something else as well. This was no fault of her own. This was done by a high-level spell which she did not possess. The only one here who she knew could and would do this was...

Sam ran out of the chapel as Joseph started awake. She gestured for him to stay.

"No matter what happens, I want you to stay in here." Joseph saw her expression of fear and knew something terrible had happened. Sam walked out to Ludicore. She walked up to his sleeping form and slugged him in the chest, which, with his stone form, did absolutely no damage.

"Wake up you bastard! Wake up!" Ludicore woke up wearing a smile.

"You require my assistance, perhaps?"

"Don't act the innocent with me, you bastard dragon! You put Sassy into a coma."

"Not really a coma. Her body is, indeed, on the edge of death, but by now her spirit should have arrived in the Underworld! If I lose... I will not be the only one going to the depths of death."

"You won't get away with this!"

"I already have. And more! I will plunder the Territories as well as this stupid valley! There is no stopping me. In fact, I'm ready to get rid of you right now. I stole her spirit and now you will join her!" he yelled. Viciously Ludicore swung an arm toward her skull with all the strength in his huge forearms. She ducked under the arm and ran in fright. Ludicore followed her through the entrance, spouting flames after her. She could feel the fire reaching behind her. Ludicore took one more breath and fired flames through both nostrils and open mouth as Sam ran for everything she had, but she knew she was already near spent. She realized that escape was impossible. Ludicore was suddenly as swift inside the caves as he was outside. There was no stopping him. She would die!

Suddenly she heard the pad of running paws and four furry forms brushed against her legs. As she turned and looked back she saw that the four animals were after Ludicore. It seemed that Joseph had made plans of his own.

Saliva passed through Ludicore's exhale and rotten stench of breath as the other pets stalked him just out of reach. The animals had effectively blocked his passage forward. Meanwhile the leader took a huge chunk of flesh out of each of the beast's hind legs. He balked and cried out in pain as the animal opened two serious wounds. He suddenly found himself immobilized. He made a final decision and suddenly vaporized into smoke and finally retreated. He formed once more with both hind legs completely intact and solidified to rest as stone with a grumble of confusion.

Saliva returned to the cave and sat near both occupants guarding them with her own life. The fact was, none of the animals had been seriously injured. Saliva's fur was smelling of smoke but none had died. That was a miracle beyond belief. From now on, it seemed, the animals would stay with her, but she did not know what to do about Sassy's dire fate.

Sam was shaken and in mild shock as she realized that her plans were a total ruin. She could not fathom what to do next. She picked up her magic book and, with much difficulty, cast her strongest Talk With Animals. This would allow continuous contact with her saviors.

"You saved my life!" Sam realized.

"From your expression I knew it was time. So I released the animals as soon as you left."

"Don't worry either of you." Saliva spoke up. "From now on we share quarters. All six of us. Ludicore has no defense against teeth and fangs. You were right. Rotten to the core!" Sam resisted a slight smile.

"What do we do about Sassy? She's in the Underworld."

"Only time will tell. I am afraid we will need help in this case. But help is definitely on the way."

"I should have never played with the Underworld. Now we have no hope."

Ludwig looked once more at the torrent of this huge river! It was so big that the other side was nowhere in sight. The Galala River was so strong that there was no question of crossing.

The troops, dismounted, sat and stared at their predicament. The crystal clear water was from the neighboring mountains. This stream seemed more like a never-ending ocean. What would they do?

"We better look for a better place to cross." Suchi suggested.

"But according to Gabrielle's maps there is no smaller section to cross. This river is relatively uniform all the way."

"There has to be a way around." Suchi insisted.

"If you know of one I am listening. But according to the maps the only place to cross is beyond Galala Strait."

"Then we go there."

"It will take time and if it is too late when we do cross..."

Suddenly everyone heard the soft sound of clapping thunder. But the sky was clear. The sound was getting louder as it approached. Ludwig turned away from the churning river and saw something in the far distance. But it was travelling at tremendous speed. As the shape came closer they made out a cream pelt... then they saw that the newcomer carried a hugh horn on the base of the forehead. Ludwig suddenly smiled with glee. The Centaur had finally shown itself. The others stared in awe as they saw the Unicorn in the near distance. The thunder slowed and quieted as the hooves of the Centaur approached. The Centaur neighed and shook its head as it stopped. The unicorn shifted, there was a displacement of space, and the unicorn took its original shape as a centaur. The being was obviously female, her breasts were ample and bare, and the troups all stared. Chestnut colored hair hung down the back of the horse-form which was colored the same cream color of the Unicorn form. Her eyes were a rich blue, the color of hazel, the color of the setting sun on the horizon. The face was young-looking with the appearance of toning and the look of glory was in her eyes. Ludwig took charge and walked to her.

"I welcome you to our presence, my lady! As you can see, we are in dire need of your guidance."

"I am here at the chosen place and time. I welcome The Guardians to my valley!" Ludwig bowed deeply. "My name is Palapany." She announced her name with the accent on the second syllable. "I am the descendant of Galala the Centaur ancestor."

"We are in dire need. We have no way to cross this river, but we must reach the other side. There are lives at stake!"

"I am the guardian and keeper of this valley. I am aware of all that happens here. I am also aware of your needs. But first we must talk!" Palapany laid down and gestured for the others to sit near her. "Samantha Littleton has lost all control over the dragon beast which calls itself Ludicore. He is now in control of the situation. He is from the Underworld. His power is great but this valley has seen many of his kind before! Samantha is frightened and has lost all hope. Ludicore expects a war, so we will gladly comply. But your power is not enough. What you need is a saint from the same Underworld, a lawful spirit who can conquer and has conquered chaotics many times before in this beautiful valley. His name is Exceptor. You will come with me to his resting place at a hidden chapel and the sword will be raised once again. Only with Exceptor's help will you defeat this beast!"

"An enchanted sword, then?" Ludwig asked.

"Exceptor rests best within a blade of pure platinum! This is the weapon which will pierce Ludicore's heart. Lucho the magician will meet us at the chapel. He is a magician... and the Exceptor's master. He alone wields the sword."

"Where is this chapel?" Suchi pondered.

"Some distance to the north. I can make the trip in two days. Your men will take three days I expect. But we must travel all day with no stops. Lucho is waiting.

"Then we go with haste."

"I agree. We must leave now."

"Mount up!" Ludwig instructed. Soon the troops, following a cream colored Unicorn, headed through the northern prairies of Galala Valley.

She was in a tornado or hurricane that kept pulling her down. She was in a constant fall that never seemed to end. She felt scared and confused, as she did not know where she was going. If anywhere. She could feel the wind in her ears and she saw nothing but the fury of her passing. Her descent was fast and furious and she found she did not breathe. But her heart was beating... or was it? Where was she going... and how did this come about? She was muddled and confused... she turned and looked down from her position. She saw a pit of blackness approaching her. Then she pierced the blackness which covered her like a blanket. It parted and she was beyond. In a world of blackness. But she saw light in the distance. But was it really there? She was falling but suddenly stopped. She righted herself and looked up. All she saw above was a wall of blackness. What she saw below was a light shade, but no one was with her. She suddenly felt lonely and cold. "Am I dead?" she asked.

Sassy could not remember a time when she had been whole. Time passed, but without the sun's guidance, she had no scale with which to measure it. Had she been here long? How long had she been traveling and how long had she been here afterward? Where was this place? And why was she here? What had gone wrong?

Sassy looked up and saw that the blackness was still there. She decided to attempt a crossing. Bracing herself she rose and burst into the blackness. But a sense of electric pain buzzed her head and sent her reeling into the light. Question...how can a spirit feel pain? And where was this place of ghosts and mystery? She recalled in her teachings, as an acolyte, there was a reference to a place called the Underworld where all dead beings were brought to await the passing. A waiting place for the dead to await their fate. A purgatory! Then she knew the truth. She was indeed in the Underworld... but she swore she could still hear her own heart beating periodically. If I'm dead then my body should be too. I wish I could find out more of what's happening out there.

[Your wish is my command.] Sassy heard the voice, but saw no one. The voice seemed to come from inside her head. Am I going crazy? She thought.

[Allow me to guide you on a journey. All that you require will be revealed.]

"Who are you?" she asked.

[That will be revealed in time. Right now it does not matter. Open your mind and allow me to lead you.] Sassy felt the tingling of a second mind melding with hers. But it suddenly hurt. [Relax. Close your eyes and open your mind.] Sassy tried to relax, and soon the alien voice was within her. She could feel his co- existence with her.

[Now open your mind to me and we will journey to the upper world. I will show you exactly what is going on.]

Sassy wondered who this was and how she knew to trust him. His voice was strong but never harsh. He spoke with a male accent... she knew this being was essentially male and she knew that she could picture him as he would be if he were human. But he simply refused to reveal his name.

[Come!] Sassy felt herself rising from dwelling in the Underworld. But then she looked back... her spirit was still there where she had been before. What part of her was it that was rising? Her mind or her spirit? [Your mind has infinate possibilities.] She felt herself pulled through the blackness with only little pain and then she was flying over the surface of Upper Earth. Her mind moved so swiftly, like the speed of wind. She could see the caves leading to Littleton... and the stone dragon in his place.

[This is Ludicore, a beast of the underworld. This is the one who did this to you.] Soon they entered the caves. Samantha Littleton lay asleep, resting on her bed covers. Joseph was rolled up in the crook of her arm. Sassy saw the animals surrounding both of them. They were all asleep.

[You understand Sam's motive for taking you hostage.]

"I understand, but I cannot forgive her."

[You must forgive! This is her toughest time ever. She has finally learned the truth. The hard way.]

"Where's my body?" she asked. The view changed and they were within the chapel. Sassy's body was where it had been all this time.

"I'm so blue. I must be dead!"

[Not dead, but you are in a coma.]

"How did this happen?"

[Ludicore cast a fifth level sleep spell.]

"Fifth level? That's too strong to survive."

[It's a miracle your body continues to hold out. But it is possible.]

"I'm going to kill all those responsible."

[Leave that to me. Come, we have one more stop.]

Sassy felt herself shimmering through the ceiling of the chapel, and then they were heading away from the mountains to the north-east at incredible speed. Sassy swore she could feel the wind in her face. The river approached with sudden speed and she gasped with surprise at the width of the waterway. Its width was more like an ocean than a river. They passed beyond to the other side.

"Where is this place?" she was inspired to ask.

[This is the land beyond the mountains. Galala Valley.]

"This is the land that Gabrielle found beyond the mountains?" It was more a statement than a question. There was no response given to her question, so she just drank in the passing beauty.

Suddenly there was firelight. A camp of men could be seen in the distance. The scene suddenly came into full view. Sassy saw a troop of horses grazing on sweet grass.

"These people are from the Territories!" Sassy suddenly realized "The camp must be three hundred strong!" She realized now just how much she was loved. "Oh, and there's Red Coat grazing over there!"

[As you can see help has already arrived. Do not despair! Soon you will be free to return to your life! But these are mortal men who must travel this earth one step at a time. You must have patience!]

They had suddenly returned to the Underworld. Sassy felt herself returning to her spirit form. Once more her soul was completely intact.

[You must trust me! And you must trust fate, no matter how hard it gets! You must trust your destiny!]

"You're leaving!"

[Trust and have faith as you always have. Everything is under my control. Sam's motives are pure. She is doing this for her own reasons. This fight is between Sam and Ludwig. Unfortunately, you are only Sam's pawn. But she would never seriously hurt her own step-sister. You must remember to trust your loved ones... and you must forgive her!]

"I'm willing to but I don't like being used."

[You must have faith in her too... she will not let you down! As I will not let you down! You must trust me! And you must trust fate no matter what! You must not give up!]

These were the voice's last words. Sassy found herself... once more... empty and alone. But the hardest part, she knew, would be the days of waiting.

Chapter 9

Ludwig Wildlife awoke with a start. He swore that moments before he had been talking with Sassy... or somebody! The message he had perceived was that Sassy was in trouble! But what could be done? He got up and approached the waking form of the Centaur.

"What's happened to my daughter?" he asked.

"You had a dream."

"I had a premonition that Sassy was in trouble. What's happened to her?"

"Sit!" Ludwig sat facing the centaur with a slight case of agitation.

"The dream was very real. I could sense that Sassy was in grave danger."

"She is... but then again, she is not. Your daughter is in the Underworld... Ludicore cast a fifth level spell and put her into a coma. Her body still lives... barely! The message you received was meant as a reassurance that she is being protected."

"How exactly is that?"

"She is protected by the spirit of Exceptor. He is able to transfer his being anywhere he pleases... he's a travelling spirit. He took her on a voyage and showed her all that has occurred. She is safe and can return from the Underworld. But first her coma must be lifted. This can only be done once Ludicore has been conquered. But he will surely try to keep her in the Underworld."

"How do we lift the coma?"

"Three fifth level dispel magics must be used simultaneously to lift this illness. With Ludicore's origins this is the only way to battle the coma. You must be cautioned, because the process is as dangerous as the illness. If the meditation is disrupted even for a second, Sassy will die!"

"I must talk to my magicians. Three fifth level dispel magics will be hard to find!"

"Lucho the magician has such a spell."

"So we need two experienced magicians from our troops who know that spell."

"And how to use it!"

Palapany concurred. "Go, talk to your men and arrange this now. If we are ready beforehand it will be much easier."

Ludwig left with apprehension because his daughter's life was at stake. As this was the third day since they had arrived at the river, and the day that they would arrive at the Chapel in the Valley as it was called by Palapany, he could feel his shoulders tensing with extreme stress. This Chapel in the Valley stood in the outskirts to the west of Galala Strait. The troops finally packed up and Ludwig called to mount. They continued their trek which would only last perhaps half that day at a moderate pace.

"Lazarus says he knows the procedure and how to use a fifth level spell. But he's never attempted three charges at once. Amanda will provide the second and Lucho has his of course. Perhaps Lucho should oversee the operation." Ludwig was explaining the situation and details to his fellow commandant, Suchi of course, who rode beside him.

"I have been reassured that Sassy is safe... for now. But I still get the chills when I think my daughter being where a person should only go after the body is gone. Brain dead! She should rightly go only once... and now there's a possibility that we might not get her back."

"You seem to be very angry with what has happened to your daughter. Would you feel the same way if it was your step?"

"I'm afraid my feelings about Samantha are mixed. I am not sure what to think of her!"

"She's a stubborn woman, but perhaps she felt this was the only way to get your attention."

"The hard way." Ludwig suddenly barked.

"What's that?"

"Children demand to learn the hard way."

"Maybe that's the best way to learn... some feel it is the only way. If only children had the wisdom of elder years. But what they lack is experience." Ludwig nodded his understanding. He suddenly blurted, changing the subject.

"I wish you would reconsider disbanding the Avengers. You guys are really good... and you have already made incredible leaps."

"We killed a man who could have been a father. He had a child, and we took him away from his family. Besides, look what happened to Sassy."

"Her capture is no one's fault. This is, and has always been a family matter. It has nothing to do with your band. It's strictly personal!"

Suchi just nodded admittance. "Sam never loved Andrew. There was no way that she could have. He was chaotic. She was lonely... and she used him to get herself some permanent companionship. That's all it is... all it can be."

Palapany suddenly beckoned and picked up her pace, soon outdistancing the others with her hooves of thunder. The Chapel in the Valley could be seen in the not-too-far distance. The centaur reached the hilly chapel and approached a form which stood out in the open air. Some minutes later the troops stopped and dismounted; approaching both human and beast.

"Greetings to the Guardians of The Gate." Lucho the magician beckoned. "As speed is our necessity we will proceed." Lucho was a chubby, graying older man, his weight outdoing his mere five foot height. Ludwig noticed that Lucho had hazel eyes and that he walked with a decided limp. He wore only a bear skin cloak which was closed around the waist. He led them around the building and entered a solid wall facing to the west. Ludwig and the Avengers followed without any trepidation.

Torch light was apparent inside the chapel, a closed in room with nothing decorating it except a platform at the far end. Lucho strode forward and opened a long chest; pulling out a sheathed sword whose blade appeared to be five feet long. Lucho then drew the sword and measured its strength. Roderick seemed spell bound by its length alone. He also noticed that the blade was made of platinum, a precious stone that also related as a currency. It had been made with all the blacksmith's heart and soul. Roderick could not resist reaching. But Lucho swung an arm out as he approached and sent him reeling to the floor in one swift motion.

"No one wield this sword except that who owns it. This is my grandfather's creation and no one touches it without consequences." Turning he addressed the room.

"Exceptor, are you available?"

[I have been waiting for you! As you know, I had other concerns to take care of.] The voice could be heard from the ceiling as a spot of bright fire glittered.

"Then take your place in the Life Sword!" Lucho commanded. The fire source streaked a line from the ceiling to the blade. Then the fire line began to shorten and it could be seen that the flame was giving the sword a life of its own. The fire line was enveloped by the blade.

[From now until Ludicore's demise I am committed to this sword. Let no one dessicrate its holiness.]

"Amen."

[Ludwig Wildlife, step forward. Your daughter is well protected. Please rest assured. From now on her fate is in your hands. I commit myself to the father of the victim. It is you who will kill the beast who captured her! Come, grasp the Life Sword.] Ludwig responded with surprise and mystery. Lucho offered him the handle proffering the blade. Ludwig nodded his assent and grasped the handle. It suddenly glowed and cast a warm, calming light which was not hot.

[I have modulated this sword to your fingerprints. From now on no one else will be able to wield the Life Sword, because they will be scorched instantly. Does everyone agree to these terms?] This took everyone by surprise. But there were nods all around. [This battle's success will depend on instant reflex. I therefore request permission to mend my mind with yours. I will control the action through instant command. Do you agree to meld?] Ludwig nodded consent. [Then it is done. Do not resist.] A bright light left the handle which Ludwig held and proceeded up his arm, flowing up to his head. Then it stopped and the glow enveloped his whole skull.

[It is done. Behold, I declare that Ludwig Wildlife is now the Master of the Life Sword. And will continue to be until Ludicore's demise. I will control bodily functions, as this allows instantaneous reflex. I have dealt with Ludicore once before. From now on Ludwig is the general who will control our battle. He is the Master. He now knows all the facts. So let him take charge.] The glow suddenly evaporated.

"I understand your commitment and now I realize my importance in the coming battle. It makes sense for me to be general."

[Thank you, master.] Ludwig suddenly raised the sword and yelled his anger.

"Peace be to this valley." he boomed. "We ride!" Ludwig lead the way out of the chapel and began to mount. The others followed suit. "Three innocents await the coming of rescue. Sam has already confessed and repented of her sins!" Ludwig faced the troops and began his war speech. "She has also given her pledge and devotion to the coming rescue. All she wants is to keep her loved ones safe! That is what she lives for now, and I do not intend to let her down! We go immediately to our chosen destination. To the battle. Ludicore awaits! In ten days' time we will do battle. And do war!" Ludwig turned and trotted off as the others kept pace with the thunder of spurring hooves.

Chapter 10

Sam felt a motion rubbing against her knees and awoke with a start. She saw that Saliva had come into the chapel to awaken her.

"Rise and shine! It's time you took your son on an excursion. You will bring him back to Village Littleton where he will be safe." Sam rose as she rubbed sleep out of her eyes.

"I know I should but Sassy..."

"You must go. I will watch her for you. Supplies are dwindling and we need to restock. Go and talk to outsiders... you need a change from this place. It's just what the doctor prescribed. You must go!"

"Alright. I could use a breather." Sam exited the chapel with Saliva right behind her.

"Joseph! Go with your mother."

"Where?"

"You do not belong here any longer. You will go back to Village Littleton where you can be safe! I don't want you here when the troops arrive."

"Are you sure I can't help?" he asked.

"You already have." Saliva replied.

"Then it is decided. I'm ready to go!"

"Go now while we still have time!"

"We'll see you later."

"Take your time. The troops will wait."

"I only hope I can be back on time."

Ludwig beckoned to Lucho who then came astride. "You have more swords like this one, don't you?"

"I have lots of platinum swords. Although they are not all like the Life Sword."

"Then we will use them."

"What are you planning?"

"How far is the river? Isn't it two days distant?"

"More like one and a half. If we keep pace we will make it tomorrow night."

"We will stop at your caves. That's where you store your swords."

"You are wise, Ludwig."

"It feels like I'm in a dream. I know everything that's ever happened in this valley as well as the history of these prairies. Will I retain that knowledge when it is all over?"

"It is a gift! To the guardians."

"I will take time tonight to finish my map. I can see every inch of this valley in my head. The river flows into the mountains in the west."

"Yes, there is another mountain range beyond the river. We call it the Galala Mountain Range."

"And to the north?"

"Unexplored territory. It is said that Galala immigrated from the north in search of food. It is not known what lays beyond this valley. No one has explored to the north. At least I haven't."

"Your ancestors have not, either."

"We've always preferred the safety of the home."

"Your wife came from Anasthasia."

"Yes, I traveled heavily when I was younger. My excursion to Anasthasia was specifically to find myself a wife."

"And she's well-bred too. The daughter of the city clerk."

"A clerk who is still well-to-do. I always swore I would marry into money. Although now all I use it for is making swords."

"Her name is Amelia. And her last name was Armitage."

"My last name is Armstrong."

"I know that."

"So we stop at my place."

"Just for the day. I have a plan that might work."

"Exceptor's idea?"

"Partially. I just thought we could rile up the beast a little."

"He does not stand a chance."

"That's true."

"You have it all planned out already."

"I'm still working on the final details."

When Samantha exited the tunnels as she approached Littleton, the sun was leaving toward the horizon. She knew that she had slept very late into the day, and the trip out of the mountains had taken several hours. Now she gestured to Joseph and picked up her pace in anticipation as she held Joseph's hand in her own.

"Home!" Joseph exclaimed. "I can't wait to have a decent meal. Rations grate on your stomach."

"I'll be back to rations soon enough I'm afraid."

"Why don't you stay with us? You don't need to be bothered by this war."

"I would if I could but I'm responsible to Sassy. I must keep her safe!"

"I think Saliva could do that just as well."

"I wish I could stay but... I don't think so. With luck I will be back in no time." Joseph shrugged his shoulders.

The town of Littleton was upon them, and Sam took a short route to the residence of Sandra and Chuck. Knocking on the door she was mildly surprised when it was Chuck who answered it.

"Come in." he welcomed. Sam trudged in and sat down thankfully. Joseph went to his assigned room. Sandra entered with wide eyes and surprise on her face. She sat down and smiled.

"I brought Joseph back to where he will be safe. I'm afraid I've been very sinful. Thinking chaos could be controlled without training. I understand now the need for defense."

"It helps prevent backlash." Sandra nurtured. "Tell me what has happened so far."

"I'm not alone in this fight. I have four pets that help defend my caves."

"And you have a three hundred man fighting force headed your way. Everything will be fine."

"I certainly hope so but Sassy's in a tight bind. Her body is in a coma and I don't know how to lift it."

"Who are they fighting? What I don't understand is why they enlisted so many soldiers."

"Ludicore is a beast of the Underworld. I found a scroll, and unfortunately, I used it."

"Now you may be more careful in the future." Sam just nodded. "Come, we have supper cooking. After a hearty meal and some sleep in a proper bed you will feel better."

"I'm sorry I ever did any of this."

"Quit fighting yourself. It's all in the past!'

Ludwig went over his map one last time. Then he pulled out a scroll which had been mapped by Gabrielle some centuries ago.

"It hasn't changed that much." he realized. "After centuries of conflict and weather it's still basically the same. The river has widened a little bit and the ice off the Black Mountains is always adding more water. But the river flows into the Galala Pass and through the Galala Mountain Range." Suchi patiently observed Ludwig's work.

"You're right. It hasn't changed that much."

"Even without traveling the entire valley, I have mapped its entirety. Only on what Exceptor gave me. It's a miracle!"

"He certainly is a help."

"Ludicore does not stand a prayer."

"Not with the both of you, General."

"I like to brag too much. I will keep my plans to myself until the proper time comes!"

"I might as well get some shut eye." Suchi stated. "Tomorrow's another day."

Ludwig went to talk to Amanda. She also had noticed the change in Ludwig, but she kept her feelings to herself. If she had not known better she would have said that her husband was not himself. He had never taken control before, and now she even noticed discrepancies in his personality. But she also noted that this behavior would be a side effect of the mind meld. Ludwig was not blind of her concerns but had put them aside for future reference.

Sassy looked up at the blackness for the millionth time and knew she could do nothing but wait. With her spare time she had listened to her own pulse which was clear and loud to her ears. It was still beating at one pulse per minute. But it remained steady and never fluctuated. By now she knew she would be safe... both above and below ground. Soon she got tired of listening and suddenly looked down. The light still beckoned. What was beyond it? In a way she did not want to know.

Ludwig fell asleep to the soothing sound of crickets under a star-filled sky. Even in his dreams he could feel Exceptor's power. He suddenly felt a tug at his mind as Exceptor disturbed his sleep and asked him to travel with him.

Ludwig felt the wind in his face as they swiftly crossed the expanse of the river. Then they approached the Black Mountains. The Beast was suddenly in his vision... unmoving and apparently unconcerned. Then they were in the caves. The chapel was occupied by a single form... a lion. Ludwig panicked for a second before Exceptor sent a sense of calmness through him. The lion was sitting near a pedistal where Sassy still lay.

[The lion is a pet.] Exceptor reassured. Then the vision changed again. Ludwig could see that they were in a strange place. But he knew what it was. The Underworld, of course. Ludwig could see the floating form of Sassy's spirit.

[We have come to keep you company, my friend! We can speak freely here.]

"Hello, love." Ludwig greeted.

"Daddy?"

"I'm here. How are you doing?"

"I'm fine, considering." Ludwig swore he could see tears in Sassy's eyes.

"You needn't worry. We are here to help you. Only, you will be here for a little while longer, I'm afraid."

"How long exactly?" Sassy asked out of pure curiosity.

"We will arrive at the caves within ten days. We have everything under control."

"I certainly hope so."

"There's no need for concern." Ludwig tried to assure her.

"My heart is beating at a low pace of one beat per minute. There is need for concern. I've been listening for it and it's still there. Although sometimes I'm forced to block it out of my mind to obtain some measure of equilibrium."

"You must realize that this is not Samantha's fault." Ludwig pressed.

"I know! I was just her pawn to draw your attention!"

"She realizes her mistake, and that she should have minded my wisdom instead of going to Littleton. She cast herself out. She's been like that ever since."

"She's certainly had a rough life."

"That she has!"

Ludwig woke early the next morning feeling refreshed and reassured of his daughter's safety. Now he could put the stone to the grinder. First thing he did was wake Amanda and describe his venturing. They talked for the final hour before sunrise. Then the troops began to break camp.

Sam awoke late the next morning and stretched her limbs with happiness. She had never felt so relaxed. A hay mattress certainly made a world of difference. Turning, she saw that her son still slept, so she got up and left him.

As Sam entered the kitchen Sandra was cleaning the wood stove. She came alongside and dug in to help get the job done more quickly.

"Tell me about Andrew." Sandra suddenly optioned.

"What do you want to know?"

"He was the child's father, wasn't he?'

"Biologically, perhaps."

"Why did you sleep with him, anyway?"

"He was a man, but he was also a creep. One creep deserves another!"

"You're not a creep. It's not your fault!"

"But I still feel guilty."

"We are getting off track." Sandra patiently reminded.

"Alright!" Samantha suddenly yelled out. "I was lonely and I stole a child so I would have permanent company."

"I know of a few men who would be glad to have you. You didn't need an outsider."

"I am an outsider."

"Only because you choose to be one. Really, you should meet a mate. I'm sure you need that as much as anybody."

"Sure I do! I used that creep. I warned him not to try for Village Wildlife. He wouldn't listen!"

"You warned him?"

"I just pleaded with him to spare himself the trouble."

"You realize their treasury is full of cleric?"

"I can imagine his face when he discovered that!" Then she made a face of utter shock.

"I'm glad you came back!" Sandra laughed.

"I am too."

The troops had ridden all that day and at dusk they noticed that the river was narrowing. Perhaps narrow enough to wade but Ludwig knew that the current would still be too strong. So he continued west until Lucho pointed out a wooden bridge which was indeed sturdy enough to cross on. Lucho lead as the troops headed across. Ludwig gestured to Suchi as he took his turn crossing.

After everyone was across they continued in the dark until they could see a light appear in the near distance. Then Lucho spurred his horse to a trot and the troops generally picked up the pace.

"This is my home!" Lucho informed as he dismounted in front of a brightly lit cavern. "Come in and meet the family." Ludwig strolled into the caves and saw that there were three living quarters. Two of the quarters were for the family. Lucho explained that the third was for visitors. He led the way deeper into the mountain to a storage room. Sitting alone in the middle of the room was a long chest with emblazoned markings on it. Luchi opened it and revealed the finely finished sheaths of platinum swords. Ludwig instructed Lucho to distribute the weapons among the troops. Then Lucho lead the way back to the central area where they would enjoy that night's meal.

Sam lugged the haversack over her right shoulder. She still felt guilty about getting Sassy into so much trouble. But she shrugged off her personal feelings and trotted on.

Chapter 11

Ludicore was tired of waiting. The troops would arrive soon enough. He knew he did not have a prayer of survival. He would be returned to the Underworld. But if he had a choice he would not be the only one staying there.

He woke up removing his stone shell and looked up at the sky. The moon was hiding behind some clouds. This pleased him. He got up from his perch and looked in the caves. All appeared dark. He knew that this was even better.

He quietly entered and slid down one of the ramps to the chamber within. All the animals were sound asleep. He then turned to enter the chapel.

At that moment Samantha arrived at the top of the stairs. She had been using her infravision to lead the way in the dark and immediately she sensed his warmth. She also made out the size of Ludicore's outline. A surge of berzerker rage suddenly shot through her. She yelled her rage as Ludicore turned to face her. She doffed her haversack and threw it across the room with all the strength in her arms and back. The bag raced across the room at full speed and landed in Ludicore's chest, knocking the wind out of him.

Before Ludicore could react she drew a silver dagger and ran to defend her sister's life. If need be, she would gladly sacrifice herself.

The move was so fast and efficient that Sam got past Ludicore's huge arms with no trouble. The dagger bit deeply into his chest. He roared his anger and pain and swung his huge right arm but Sam tucked and rolled out of his reach. Then the animals were on him! Saliva came from the chapel and bit into his leg tearing a huge chunk from one knee. Ludicore screamed his anger once more and then took his retreat as red smoke.

Samantha suddenly began to shake as adrenalin shot through her system. She had defeated the beast all on her own. And she was alive to talk about it!

"We never expected it!" Saliva apologized. "He's really desperate!"

"He won't try again... I hope. It's lucky I arrived when I did!"

"What a monster he is!"

"He wants Sassy to remain in the Underworld. Maybe he hopes to torture her."

"She won't remain there for long. Either she returns or she passes on. The Underworld does not work that way!"

"That's what I'm hoping too." Samantha suddenly noticed that her whole body was shaking with fear and anger. "How dare he?"

"Try and get some sleep." Saliva suggested. But Sam ignored her and entered the chapel. Sassy was still asleep and safe in her coma. Sam felt relief surge through her bones as she grasped Sassy's hand.

"I'm sorry! I'm so sorry!" she cried her regret. "I'm sorry I put you through this! I'm sorry!" Tears cascaded down Sam's face as she kissed Sassy's calm cheek and then she sat and watched, listening to the pattern of Sassy's steady breathing. That sound nurtured her to sleep.

Five days passed before Sam felt calm enough to relax. She had eaten and slept the whole time. But now she was sure that the beast had tried his best. She only prayed that he wouldn't try again.

At that moment the troops were approaching the eastern border and the Black Mountain Range. They had followed the flowing river and knew that they would arrive before nightfall that day. Five days had passed since they left Lucho's habitat. But Ludwig knew that tonight they would do war!

Ludwig called a halt as a white unicorn approached. Palapany was right on schedule. She entered the herd and they continued on. In mid afternoon they arrived at the mountain range. Then they turned south toward the caves.

Sam suddenly awoke to the sound of horse hooves. She immediately went out and saw a thundering cloud of dust as the troops approached. Ludicore woke up on cue. Lucho was the first to approach the beast.

"Hello, Ludicore!" The beast nodded his head in reply. "You realize that we are here to confiscate your prisoners!"

"To get them you will have to kill me!" In response Lucho drew a platinum sword.

"To your death." he yelled. Ludicore swung his tail and cut the sword in half at mid-blade.

"I've learned a few things since last time. You'll find I'm full of extra surprises." Ludicore gloated but Lucho just turned and retreated as three riders rode up and drew platinum blades. This confused the beast and in a rage he broke all three swords with one stroke. Then three more drew platinum swords and he suddenly exploded in a rage of fury. Finally Exceptor spoke.

[You cannot destroy what you cannot see. You are finished!] The voice seemed to come from everywhere. Ludicore had nowhere to turn. So he decided to take the offensive.

It was reaching dusk as Ludicore looked down and stared steadily at nothing but earth. Suddenly two lightning bolts sizzled out of his eyes and struck the ground in front of him.

The ground began to shake and a huge quake formed into a hole twenty meters wide and five feet wide. Lucho took the time to inspect the crater.

The hole was opening to the Underworld, he knew. As Lucho watched, beings began to come out and attack the troops. Ludwig recognized black vampire bats. Some mummies also appeared.

The land became a mass of swordsmen and sorcerers fighting for their lives. Mummies were easily taken out by fireballs. But the bats were quickly becoming a huge problem. Ludwig took this all in at a glance and then acted. Pointing his Life Sword at the sky, he released a surge of energy which shot to the heavens. Two minutes later a tornado formed.

This power from the Underworld touched down and collected all the bats and the corpses of the mummies. But it never touched the humans. The creatures spun to the eye of the tornado as it disappeared down the crater.

Ludicore had been taken by surprise and was shocked at the sight of the Life Sword in a stranger's hands. He raged with anger and swung his tail one last time. But, instead of breaking the sword, the tail was suddenly cut off at the base.

The dragon yelled with rage at the pain and finally realized that he was done for. Ludwig wasted no more time. He raised the sword and yelled his triumph.

[It is time! Release me!] Ludwig felt Exceptor release control of his mind as a surge of energy leapt into the sword. Then before Ludwig knew what happened the sword was firmly planted in the beast's chest where his heart would be. The beast took one deep breath and then fell in the hole head first.

The crater then closed leaving no sign of turned earth. The land was whole once more. But before the crack sealed a platinum sword rose from the crack to a laying position on green grass. Ludwig picked up the sword and raised it in victory.

Sam was suddenly running toward Ludwig and screaming her happiness. She hugged him to herself with joy. "That was some battle! Wow! Talk about wasted! I was sure he would break your sword!" Ludwig hugged her back with much affection.

"I'm sorry! I guess you just had to learn the hard way!" Sam suddenly shrugged with worry.

"We still have Sassy!" she reminded.

Sassy suddenly saw the blackness change to clear sky. Then two beings appeared. One, she knew, would be the dragon. It had finally come! The dragon was closer to her than the man and was headed straight for her. Then the man sent a shock of lightning through Ludicore's back, sending the beast passed Sassy and further into the light. The man approached with a smile.

[Hello, my friend! This is the real me!] Sassy took a good look at the man she was with. He had brown hair, stood at six feet looking down on her, and was groomed with a clean shaven face and a pudgy nose. His eyes were a hazel that Sassy had never seen before. He wore a robe of pure white cloth.

Lucho stood at Sassy's head and saw that she was in bad condition. They would be lucky if she lived. The sorcerers announced that they were ready. It would take three fifth level dispel magics simultaneously to revive Sassy. Unfortunately, it was a risk they had to take.

Lucho stood at Sassy's head, Amanda beside him and Lazarus stood at Sassy's feet. Then they all began to speak archaic words. Force could be seen to come out of their finger tips and slowly a field of energy began to form. The magicians began to move their arms and each one went from head to toe generating an energy field vaporized instantly. But the field from their hands had to be maintained continuously. They reached Sassy's toes and began the return trip to her head. They went back and forth two more times and then removed themselves from her place. Now they could only watch and wait. Tralina pushed in and felt for a pulse. There was no response at first. But as she kept trying she began to feel pulses thirty seconds apart. Then they began to increase and at the last her heart was beating at a steady 45 beats per minute. Tralina smiled her happiness and satisfaction that the magicians had done their job properly. It would only be a matter of time before she recovered.

[It's time you go!] Exceptor reminded. Sassy was filled with happiness at her new-found freedom. She hugged her friend and a tear suddenly fell from her face.

"Your name! I have to know your name!"

[As a spirit I am known as The Exceptor. But in life I was a human magic user. After I died and passed on I decided to return to the Underworld to help others who are in passing. There are others who do the same thing as I. I am the spirit in charge of Galala Valley. All the dead come here. And the dying. My name was Arthur Lenningham of Anasthasia. But now I am only know as The Exceptor. You are the only one who knows my real name now. So please keep it to yourself!]

"Why are you telling me all this? It's frankly none of my bees' wax."

[Because you are the first person I have encountered that did not belong in the Underworld. I have taken care of that and now it is time you returned to your own life!]

"Thank you! For everything. I can never repay you."

[It's all in a day's work. No thanks are necessary. It's just my job. Now go. Your family is waiting for you.]

Sassy waved goodbye as she rose to the sky above. And then she was back.

Sassy awoke peacefully. She sat up and stretched her body to get the feel back and the blood running again. Then she stood up. She suddenly noticed that everyone in the chapel was staring with wonder.

"What?" she complained.

"You went to the Underworld. How did it feel?" someone asked. Sassy just shrugged.

"I can't explain it any better than Exceptor could. So ask him. Boy it's good to be on my own two feet again." Sam approached and put her arms around Sassy's waist.

"I'm sorry. I'm just glad that's all over."

"All is forgiven." Sassy responded. Suddenly she spotted a familiar figure in the room from long ago. She pushed her way through and approached Alvin Littleton. She just wrapped her arms around him and planted a huge kiss on his lips. Then she smiled at him. "Long time no see." Alvin just nodded. "Are you through with the Elven Guard stuff now?" Alvin nodded his head.

"I'm all yours." he responded. Sassy laughed her happiness.

Tralina approached and wrapped her arms around Sassy. "Hello, cous. It looks like you're doing fine. I'm sorry I wasn't there for the wedding. But we have the future. And I have my future back. So everything's turned out fine." Tralina was in shock. How could Sassy know all this?

"You knew?" she asked.

"Yes, I knew that my mother was your aunt. So that made us cousins. I only discovered it the day the thieves broke up the party. I was planning to tell you that same day."

"Well..." Tralina was in shock with surprise. "I was going to spill the beans to you. I thought you didn't know."

Ludwig approached Sassy as the chapel cleared. "I want my two lovely daughters to walk the short route with me. Through the caves." The group of family members cleared the chapel last and immediately left down the stairs to the caves. The troops headed through the darkness of Galala Valley as the Wildlife family took a shortcut. The Elven Territories were safe once more.

Samson's Return The Epic Biography Of Samson Trellis

Samson Leader is a Dwarven guide... and secretly, a prince. When three brothers of his own kind show up in his home town he has a big surprise coming!!

The Dwarven Clan Wars have called a seize fire... and these three young Dwarves have come to bring the lost prince home. After twenty years of bloodshed, Samson Trellis, a.k.a. Samson Leader, is the only one who can put a stop to the war completely.

Samson's Return is the epic sequel to The Avengers Trilogy. Samson's Return is written by Sassy Wildfire and Carolyn Christine Lastiwka. The trilogy is composed of The Avengers, book 1, The Eight Avengers, book 2, and The Wildlife Campaign, the conclusion. Sassy lives in Edmonton, Alberta, Canada, where she lives and writes as a pass time.

To Crown A King Pig's Cleft Inn

Three Dwaven brothers slowly entered the Pig's Cleft Inn. The one in the front looked across the room. He felt somewhat uncomfortable due to the very nature of their assigned mission. His name was Simon. He looked behind him unsurely, at his two younger brothers, Walter and Timothy. As the eldest he knew that he himself was commander-in-chief of their assignment. Their current mission was to look for, find and bring home the Legendary Prince of the Darven Territories, Samson Trellis. Simon suddenly decided to give up any form of secrecy he had thought they would need and addressed the whole room.

"Does anyone know of a Dwarven personage known as Samson?" Simon knew that the first name would be enough. The room was suddenly very silent. Many of the customers turned to stare at the newcomers. Simon took the sudden silence to mean that Samson, whoever and wherever he was, had made many friends! A barkeep, which stood at the bar, gestured for the brothers to approach. The man leaned against the bar to draw himself closer.

"My name's Amin Nanderson. I know Samson... very well! He calls himself Samson Leader! He has always been considered a highly respected member of society. He works as a travel guide, and I'll tell you how good he is, he's never lost a single client!"

"How old is this Samson?" Simon asked.

"He's just celebrated his 42nd birthday! November 25th. In Dwarf standards he's middle aged." Simon turned to his brothers to confer.

"Our Samson was 22 years of age when he left! The time since and the war has covered 20 years! That means that our royal would be exactly the same age! I'm sure that this Samson Leader is indeed our man! Now, we have the identity, but how do we find this Samson Leader?" Simon asked. He stood leaning against the bar and Amin poked his shoulder from behind. Simon turned.

"I'm not done... not by a long shot!" Amin confessed. "If you want to find Samson go up river! His home town is called Colina, it's the very next town up!"

Simon stood staring at Amin in stunned fascination. This mission was, in his mind, coming to fusion with drastic speed! He suddenly needed to sit down... he grabbed a table. He folded his knuckles as his younger brothers sat.

"As leader of this campaign I'm finding that this mission is already coming to fusion! We just got to this ocean front and already we have a name and a location!" Simon shook with physical fear. His eyes teared. "Our high prince is indeed still alive... and he is also well-known under the pseudonym of Samson Leader!"

"How do we know that this is really the same man?" Timothy asked.

"We don't!" Simon confessed. "My fear is, if he's not, we'll be the laughing stock of the whole city!" Walter was also thinking.

"The legend says that the prince has proof of his lineage... the Trellis Family Crest."

"This Samson Leader is the only Dwarven personage that Amin's ever heard of in these parts! And he knows Samson well!"

"Should we trust him?" Timothy asked.

"I feel we have no choice!" Simon responded. Amin suddenly approached their table with three tumblers of ale.

"On the house." he offered. He set the tumblers down and grabbed a seat of his own.

"If you boys are looking for your prince you won't probably find him here! Anasthasia's an ocean front... he comes here only when he has an important contract! When he has to!" Simon gawked at Amin.

"What you're saying is..." Simon shed tears of uncertainty.

"Samson Leader is certainly the same man... Samson Trellis!"

"You're sure about this?"

"Aren't you?" Amin asked. "One thing Samson always swore... when he finally did claim his title the first thing he would do is find himself a young wife!" Simon freely acknowledged that fact.

"Despite our differences we do indeed have drive! The Legend of Samson Trellis is coming true!"

"His story was made into a legend?"

"Our father spread the legend to give faith and belief to the whole nation... it was a promise... that some day Samson Trellis would indeed come home!"

"You boys have been assigned to find this long-lost prince!"

"And apparently we have!"

"Samson Leader is the only Dwarf who's ever been a prominent citizen of this nation! I'm sure he does, indeed, miss his homeland!"

"How do we get to this... Colina?" Amin thought of something.

"The Elven Guard's just pulled into port! Go with them! Your travelling time would be cut in half!"

"Elven Guard?"

"It's the equal of The Avengers. They're the policing force which travels the rivers. I used to be a member... until I decided to retire and set up shop! Colina is where you'll likely find him! His regular pit stop is called the Boar's Head Inn. Sometimes he and his fellow Avengers used to use it as a drinking place!"

"Avengers?" Simon asked.

"News about local politics." Amin laughed. "The Avengers used to be a government-run policing force for the whole principality! They have been associated with such dignitaries as Sasselia and Ludwig Wildlife of the Elven Territories! And Samson Leader is well-known as the Dwarven member of that band!" Simon was breathless with the prince's military experience.

"Has he been in many battles?" Walter asked.

"Samson Leader is one of the most honorable and learned fighting men in known land! He likes to brag about his accomplishments... especially when he's loaded. But I'll tell you one thing... when he's feeling good... on booze... he opens up... bleeds his heart! That weakness is what led to the revelation. Now tell me... is the war really over?"

"The Dwarven Council has called a seize fire!" Simon assured.

"A seize fire... but it can't last!"

"The war will only be officially cancelled upon the return of an heir!"

"Samson is drastically needed at home!" Amin confirmed.

"Why do you help us so?" Simon asked in the dignified tongue. "It was without need."

"And let the prince continue to wart?" Amin answered. Simon smiled wide with Amin's knowledge of the language.

"I humbly request continuance."

"Samson Leader is a very lonely man... and I know for a fact that he's always been chaste... whom in this world could he ever find to mate? He's as strongly driven as any man I've known. But his loss has led to bachelorhood... he's like any man... deserves a proud wife... and infants of his own blood! Everything his own realm would provide."

"We, as the Waterbee Trio, have vowed on our future king's own life, to bring him home to all such things. Infants of his blood, matrimony to a fair laid maiden, and the glory of his own rule!"

"Your prince is considered the Prince son of Anasthasius himself! He has been honored, along with his Avenger counterparts, as one of the most beloved and powerful forces in this whole nation and principality!"

"I ask again, why did you help us so? And why didst thou keep silence all this time?"

"Because, my friend, he is one of the few men who are truly powerful! And because the barkeep is the drunk's best counsellor!" Amin sat crying the happiness of peace. Samson was indeed going home!

Boar's Head Inn

Three days had passed. The Elven Guard approached Colina Waterfront. Three Dwarven brothers stepped off onto the warf. A stranger bumped into Timothy and then passed. Simon observed him closely from a distance.

"Who or what was that?" Timothy wondered.

"Thief!" Simon answered.

"What?"

"Checking for currency. Pick pocket!"

"How?"

"Neutral, obviously!"

"Why you say that?"

"Only chaotics backstab!"

"Backstab?" Timothy gasped.

"Do you have currency on you?"

"Father demanded that you carry it all!"

"Walter?" His brother held empty hands. "Good, I'm carrying it all!"

"Backstab?"

"Let's get out of this filth before we're all slaughtered." Simon suggested.

Following the directions Amin had provided they soon arrived at the Boar's Head Inn. They entered and chose a table at the back.

"What do we do?" Timothy asked.

"Wait!" Simon responded.

"How long?" Walter asked.

"As long as necessary!"

"Shall we get a room?" Walter asked. Simon noticed the barkeep staring at them. He approached around the bar.

"Don't get many Dwarves in my bar!" he revealed. "What'll you have?"

"Three ales!"

"Pay me before you leave!"

"You're very kind!" The barkeep turned into the back. A young human boy, his son obviously, went running out the front door. Simon cursed and hit the table. Their purpose here was too obvious! The barkeep returned to their table.

"My son's on his way to find your friend! Samson, I believe!" Simon shook his head in frustrated assent.

"We were hoping to surprise him!"

"And you will! The message was to come... I have further news about the war!"

"I didn't catch your name!"

"Peter... Huntingham."

"I assume you have inside contacts?"

"Many!"

"Does Samson know?"

"About the seize fire, yes!"

"That's the latest so far!"

"I should certainly hope so! I don't like to be dysfunctional in my duty!"

"I expect he's been watching!"

"Ever so closely!" The boy came running back in.

"My lord Samson comes!" he assured. Simon's heart fluttered. Peter stood behind the bar once more. Samson came in at a rapid pace and did not bother to look at the back of the room.

"You said you had news?" Simon heard in the distance. Peter then pointed toward their table. Samson turned where he stood. Then he literally stared. Simon and Walter both rose. Simon asked Timothy to stay put. The brothers sat on either side of Samson at the bar. Samson seemed confused with their presence... and a little concerned.

"What do you want from me?" he asked with an edge to his voice. Simon spoke up.

"Have you ever been acquainted with a clan named Waterbee?" Samson thought. He stood silent for two whole minutes. Then his eyes began to tear. He nodded his head forcefully in confirmation. Simon padded Samson's back comfortingly. "My lord, we have come to bring you home!" Samson recovered somewhat, clearing his eyes of tears. He turned to Simon and literally stared. He suddenly noticed his apparent indiscretion.

"Forgive me!" he apologized.

"Stare all you like!" Simon assured him. "You've earned it! I tell you one thing, when you lay eyes on the first Dwarven maiden back home you will stare!" Samson both laughed and cried. "I pass my leadership on to you, my friend! You are now the leader of the Waterbee Trio."

"I require your names."

"I'm Simon!"

"I'm Walter!"

"And that back there is the baby... Timothy!"

"Simon!" Samson remembered. When I left you weren't even born yet!"

"You remember!"

"Your father swore he would name his first-born son after himself! Simon Jr."

"That's me!" Samson hyperventilated with joy. He wiped away more tears.

"I cry for love of my homeland... but I also cry for all the wasted years! And so many lives... gone! Simply because of Clan Honor!"

"I must warn you..." Simon confessed, "the land is not the same. The forests from your home to Clan Waterfoot are all ash!" Samson nodded his knowledge.

"But the land will indeed recover! I estimate that Clan Trellis will already be green! My father's equipment still stands, right where we left it!"

"The only man who could melt platinum. That was Andrew Trellis."

"He was the best... after a while he became the only smithy in the Territories! People came from everywhere to enjoy his service! He gave me my family crest when I killed a roe deer... my first kill... I was sixteen!"

"You indeed are a skilled archer!"

"The arrows I carried had silver tips. My first cast went right through its fleshy neck. It never saw it coming."

"The first kill is indeed the most important." Samson removed his dagger from its sheath at his waist. It glistened across the room. Timothy was instantly across the room. Samson laid the object in his open hands. The brothers looked closely. This dagger had always been Samson's trademark. Simon picked it out of his hands.

"Pure platinum."

"This must be worth a fortune!" Walter wondered. Then Simon gazed at the crest emblazoned on the handle.

"The Eagle!"

"My family crest! My grandfather was the owner of a pet falcon! This was his crest. But my father wanted his own crest! The eagle!"

Each of the Waterbees got to inspect this finely crafted platinum weapon. Samson soon returned it to its sheath.

"How soon can you leave?" Simon asked.

"Immediately, if need be!"

"We should decide on a route."

"The shortest!" Samson exclaimed.

"Straight line?"

"Across the river and directly south-east!"

"Is there a way to cross?"

"By ferry! And we need riding horses!"

"You'll take care of that, I presume!"

"I do have access to both riding and war horses!"

"You're an Avenger, after all!" Samson smiled his mild surprise.

"You have been checking up on me." he accused.

"Amin told me the whole story!"

"You've been to Anasthasia?"

"That was our first stop!"

"Pig's Cleft Inn. It's been twenty years and my barkeeps have always kept it to themselves. Your barkeep is indeed your best friend! Give me two days! I need to make arrangements!"

\Ambush At The Border

The Waterbee Trio and their royal host had finally arrived at the border to the Dwarven Territories. Samson clearly recalled this route from when he had left the Territories at 22. Simon Sr.'s men had left him at this very spot to continue on his own. He finally stood on home soil.

"Let's set camp!" Simon suggested. Samson shook the negative.

"Let's continue on to the closest settlement to seek shelter."

"It's dark! We can't see!"

"I think it's worth trying!" Samson assured him. Simon nodded in reluctance.

"The troop continued, crossing mostly grassland. They then reached the outskirts of deep, penetrating forest. Samson moved on, with Simon following with some concern. But then again, Samson probably knew what he was doing. They traveled for another fifteen minutes before they saw the flames of a distant fire. They were indeed close.

Samson felt a sudden crack of pain on his skull from above. Everything suddenly went black. He suddenly fell.

Myna Milner sat sewing a blouse. Aged at 22, she had yet to marry. She had refused the promise twice in the past four years. Myna was different... she didn't want just anyone for her husband. She visioned an older, experienced husband. He must catch her eye! One that stood as her mental equal. He also must match her own maturity. Most of all, an older man who's never wed.

Three men came in lugging four unconscious forms of strangers. They dropped the bodies roughly to the floor. The victims' hands were tied behind their backs.

Myna became immediately furious in the face and concerned at the same time. Her mother, Alanah, came in. She saw the prone forms lined up on the floor. She suddenly slapped her oldest son across his scalp.

"Why?" she asked.

"They invaded private property."

"Sonar! When are you going to learn? The war is over! You can't ambush strangers just because they're passing through!" Myna went directly to work. She was trained as a healer! She had been taught by her own mother.

She gently cut the ties on each and then turned them on their backs releasing their arms so the blood could return. Then she checked the pulses. Then she got a good look at them. She suddenly gasped in surprise.

"Mother!" she called. Alanah immediately came over.

"You know who they are?"

"This one is Simon Waterbee! Sonar ambushed the Waterbee Trio! We'll never live this down!" Myna had tears in her eyes.

"We'll give our apologies, of course!"

"Let's just hope this doesn't lead anywhere. Simon, Walter and little Timothy. All three!" Myna turned toward the fourth victim.

She saw a few wrinkles and a greying beard... she suddenly took extra interest. This man appeared older than the others... but she had no idea who he could be! She went to arrange for a hot cloth on his forehead. When she returned she got very curious as to the stranger's identity. He certainly qualified as the ideal mate! Would it wake him to... she lifted one of his eyelids. The irises were blue... her own eyes were a constant shade of chestnut brown. Would allow for variation in eye color!

She suddenly heard a whine from one of the others. She turned... Simon was now coming to. She faced him and tried to reassure her patient.

Simon opened his eyes. He tried to spring up. Myna gently held him down as Simon clasped his head in pain.

"Lay still, Simon Waterbee! The nausea will pass!" Simon struggled knowing that Samson's safety was more important than his own. He recognized his location as the Milner residence. After five minutes he was well recovered. He tried sitting up gently. He looked around noticing Myna in attendance with deep concern.

"Hello, Myna." Simon greeted. "I humbly thank you for offering us shelter!" Myna winced with the sarcasm.

"Who is he?" she asked. Walter was coming around. Then, five minutes after him, Timothy followed. Myna knew that her questions would be forced to wait until all three had recovered.

Samson was still out. Myna continued nursing. Whoever this stranger was she knew that this would be her only chance! The sudden knowledge coursed through her bones. She knew that if she couldn't have this man her life would be ruined. She turned to the three brothers.

"He's very mature and aged... but he's never been seen before! At least not in these parts!" Simon knew the truth would be best! But how did he tell her? Timothy and Walter remained obediently silent. "Simon, who is he? Frankly, I think he's an outsider! Although he is very much one of us! Your troop was coming from the border line."

"He is new to the Territories."

"How old is he? He can't be much more than forty!"

"He's currently forty-two." Simon revealed.

"Twice my age and yet no older!"

"Myna?" She made her decision.

"I want to ask for this man's promise!" she revealed. All the boys suddenly shook with obvious confusion... they all looked at each other. How could they tell her that if the promise was kept she would become the next queen?

Their confusion was short-lived. Samson suddenly recovered. Myna ran up to him and waited within sight. She was the first person Samson laid eyes on.

Samson's vision began blurry. He sat on his elbows gently. But he recovered quickly without any nausea. Then he laid eyes on Myna.

Myna Milner had brown eyes... but she also had blonde hair and eyebrows. Her nose was pudgy and her lips were thinly made. As with all Dwarves her ears were pointed. Samson placed a hand under Myna's shoulder length hair and pulled his hand to the ends. He gazed at this young woman for a long time. Myna was stunned. His attraction left her breathless. Samson spoke.

"You'll have to forgive my asking, but which settlement is this?"

"The Milner settlement." Alanah offered. The Waterbees stood silent with suspense. Would Samson reveal himself?

"Boy!" he called. The three culprits appeared. He pointed to the eldest. "I understand why you did this! There is much insecurity in this land! You will go ahead of me! In the morning you will travel to each and every settlement and village in the known Territories. You will spread the message that Samson Trellis has returned!"

Simon physically laughed with glee. Sonar turned white with surprise. Myna cried tears of frustration and began to pace. Alanah's face was also white. She sat herself down and put her hands to her face.

"The legend states that Somson will prove himself to all! I carry my family crest." Samson removed his dagger from its sheath and offered it to the lady of the house. She peered at it through open fingers. Samson laid the dagger down if front of her for the offering. No one dared touch it!

"Your crest is the Eagle!"

"Or the Falcon!" Samson offered. Myna continued to pace, crying physical tears of grief. Samson looked at her and spoke. "Talk to me!"

"How can I ever ask you for your promise?" she wondered. "The prince himself."

"I am only an old man!" Samson offered. "I am nothing by myself!"

"Your title..." Myna reminded.

"Is irrelevant!" Samson insisted. "I am only a humble old man!"

"You act very meek!" Alanah suggested.

"It was my father's gloating pride which led to the war!"

"You have never married?"

"Who is there to marry outside of the Territories?" Myna had finally somewhat recovered.

"What do you expect from your mate?" she asked. "The future queen."

"She will be my counselor, my mate and my equal."

"How can I ever be your equal?"

"I am old... my title gives me more status than I deserve! I think that you are already my equal. You hold all the requirements. What's your name, by the way?"

"Myna... Myna Milner!"

"Well, Myna Milner, you are young enough to carry safely but also appear old enough to be committed!"

"I'm 22 years of age!" she informed.

"You will find that Myna is very mature for her age!"

"Comes from training!" Samson suggested. "And hard life lived wisely!"

"I'll make sure Sonar does as you ask!" Alanah assured.

"The message must go before me! My parents were my counselors. My teachers. And my loved ones. But they are both gone now! I will be forced to depend on my own mate to level me out. Whomever she will be!" Myna stood shocked at his obvious impersonal tone.

"Your father's name was..."

"Andrew."

"And your mother is Aleesa."

"Aleesa's father was named Samuel. Samuel Weatherbee." Simon suddenly looked confused.

"Samuel Weatherbee was your grandfather?"

"My mother's father, yes."

"How can that be?" Simon gasped. Tears began to fall from his face.

"Simon, my friend, what is it?" Samson wondered.

"Samuel Weatherbee is my uncle. Simon Sr.'s brother." Simon revealed.

"What?"

"It's true." Alanah confirmed. "Samuel changed his name when he married... he started a whole new clan and built his own settlement."

"Then the Waterbees are my second cousins!" Samson declared.

Anders Settlement

Samson stood adjusting his horse's saddle. He did not yet know what his next step would be. But he would stay at Milner Settlement until he knew what to do next. Alanah was heard approaching from behind.

"What will you do now?" she asked.

"I don't know yet." Samson honestly answered.

"When you mentioned Aleesa I remembered a small rumor I had heard years ago! So I sent my scribe to go for the Historian. He's a well-known Clan soldier which was assigned to document war statistics. Casualties, deaths, assault movements on both sides. He also documented settlement raids." Samson continued to play with the saddle but Alanah knew that he was paying very close attention.

"He told me the story of the raid on Clan Trellis."

"Both my parents were killed in that raid!" Samson reminded. "The only reason I survived was through the honor of Uncle Simon!" Alanah decided to forget the details and just get to the point.

"Samson, Aleesa still lives!" Samson immediately stopped his struggling and turned to her.

"You're telling me..." he opened. Alanah just nodded her head. "I want all the details!" he ordered. He began to walk. She followed.

"Simon picked you up onto his horse and left your mother suffering! He couldn't be sure that your mother was gone! But he could only carry one! One of Simon's fellow fighters was a man named Satira Anders. Both were on your father's side. Satira was at the raid. He saw you get onto Simon's horse. The raid was nearing its end. Most of the troops were chasing the enemy. But, Anders said later, something had stopped him. He had caught a movement off in the corner of his sight. He looked down... he saw one of the bodies moving. He got off his horse and turned the body over to check for wounds. The arrow had taken her in the waist and abdomen. The healer said that the abdominal bones had reflected most of the damage. Anders checked for a pulse. She was still very much alive!"

"He took my mother to safety!" Samson concluded. "I owe Satira Anders a debt of gratitude!"

"There's more!" Samson gave up in confusion. He turned to face her. "Satira Anders later became her husband! They have two offspring. A boy and a girl!" Samson grabbed his forehead in confusion and wiped his face.

"So not only have I found distant cousins, I also have half siblings!"

"Samson, since the war a lot has happened. And history has changed with the times."

"I'm very well aware of the theory of time lines."

"I'm only trying to help."

"And I appreciate it! I really do!"

"How you feeling?"

"Somewhat confused!" Samson confessed. "My mother would be 62 today."

"Will you go to Anders Settlement?" Samson nodded reassurance.

"Of course I will! I guess that's my next stop!"

"Sonar's gone ahead... his first stop is Anders Settlement. His message is to expect an honorable dignitary to make a visit some time soon. And I insisted that he talk directly with Aleesa!"

"So she will be expecting somebody!"

"But she won't know exactly who!"

Some time later Samson returned to his saddle. He wanted to pack his saddlebags... he decided he would leave the next morning. Myna suddenly intruded on his concentration.

"My lord?"

"Myna, that title is totally inadequate."

"Inappropriate." she responded.

"My name is Samson... so use it!" he reminded.

"I hate to ask but... you never confirmed the promise!" Samson stopped what he was doing, took her by the hand and began to walk with her.

"I'm afraid I can't make the promise official... much has happened and I have many places to go! Things are still too unsettled!"

"Too unsettled to devote yourself?"

"Right now my people need me. My kingdom needs me." he answered gently.

"There's no way I can help?" she asked.

'Yes there is... by being patient! If I don't come back or call you within six months you may come to me."

"Is that a promise?"

"Indeed, it is!"

"Then you have made the promise!'

"In a skewered sense, yes!"

"I guess that's the most I can ask!"

"I will send my scribe to see you often. I will always tell you where I am. If we are to wed you have that right!"

"You are committing yourself!"

"Committing and promising are two different things!"

"How you mean?"

"Promises are too often broken. Committing is a process of follow through that guarantees success!"

"So making promises is not such a good idea!"

"I never promise anything... period. It's against my very nature!"

"That's why you hesitate!"

"I can't make a promise but I can commit."

"You're bright. Wish I had known that! When do you leave? Not too soon, I hope."

"I leave in the morning."

"For where?"

"Anders Settlement. My mother is the lady of the house."

"I'm so happy for you! Where you going after that?"

"I will bring the Waterbees home."

"And have a long talk with Simon Sr.?"

"I certainly hope so! After that I will go to prepare a home!"

"For both of us?"

"And our children!" Myna blushed.

"Where will you build?"

"Right where I always have... Clan Trellis!"

"So what will you do there?"

"Same thing my father did!"

"You're a smithy?"

"My father taught me for four years. I could do any general job that he could!"

"And melting platinum?"

"I'm afraid that's a process that may have died with my father!"

"He never taught you?"

"He started to.... When I was 21. But before we knew it the war had come! Then all thought of platinum smithing was wiped out forever!

A Presence With The Queen

Samson and the Waterbee Trio approached Anders Settlement. When they reached the villa Samson called to dismount. Walter offered to settle the horses. He took Timothy with him.

"You coming?" Samson asked of Simon.

"How long you plan to stay?"

"I think we might spend the night!"

"As you wish." Samson headed directly toward the house. Sonar met him at the door.

"You haven't left?"

"Her Majesty insisted I stay! She will not allow me her leave until she gets her answers." Samson smiled his knowledge. The queen, his mother, had never liked mysteries. They usually meant that something remained hidden.

Samson entered and stepped directly into the living room. Simon stepped up to join him at his side. Satira Anders turned to face them and immediately recognized Simon. He stepped up and shook his greeting. Simon then introduced Samson as the dignitary the queen had been told to expect. Satira confessed that Aleesa had agreed to a private audience.

"Despite her frustration she does know how to accept guests!" He led Samson to a door. "This is my private office! I'm afraid the queen has made it her own!"

Samson stepped in and the door silently closed. He stood facing the door... he didn't want to face his mother and possibly be accused as an imposter! So he developed a plan. He turned to face his adversary.

Aleesa Trellis Anders was 62 years of age, to Samson's own calculation, but as soon as he looked at her he had to smile. So many memories of home! He shed one tear... he simply wiped it clear.

The queen's eyes were still his mother's. Hazel green irises, dark pupils and diamond shaped eyelids. Her hair had grey, but not, amazingly, much. Most of it was still predominantly black. It still hung folded back in front and the back flowing beyond the shoulders. He noticed that her skin was still predominantly wrinkled and fragmented... which always came from being a smithy's wife. He held back sudden tears of love and turned to royal courage. He knelt before her feet.

"I greet you, your majesty!" Aleesa laid a hand on Samson's scalp, only half heartedly, noting that the atmosphere of the ceremony had been broken.

"Your delay is considered inappropriate."

"Forgive me, my lady, but I hesitated out of fear!"

"Fear of me?"

"Yes!" he confessed.

"If you're a dignitary than there's no need to fear your equal!" Samson smiled with the irony of her statement.

"That is something I tried to tell someone else... but I failed to state it so clearly."

"How can I help you?"

"I have come to take my place at the throne to this kingdom! With your blessing, of course!" Aleesa stood stunned and felt a little bit defensive. Her reply was obvious, Samson knew. Aleesa spoke in anger.

"I'm afraid that the only man who could or would rule in my place is my own first-born son, Samson Trellis!" Samson stood on the verge of loving tears. "Unfortunately, I fear his demise." Samson's response was clear cut and obvious. But he used the extended tense in his response.

"He also feared your demise, I assure you! When you fell he was right next to you! It could as easily been he who fell. I assure you, he had very similar fears."

"You are well trained in the dignified tongue. Are you a historian?" Samson thought quickly. His past stood full of clues which would clue her in to his true identity. He congeculated how long her lost faith, now obvious to his eyes, would blind her.

"I was born and bred a smithy." That was his first clue. Aleesa seemed somewhat surprised to find a fellow blacksmith.

"If you please, I would be very interested in knowing your particular element."

"I deal with..." Samson paused. He gave her one last clue. "Platinum weapons."

"The only man I ever knew that could even melt the element was my own equal... the king of these parts!" Aleesa sat on the very edge of frustration. "Surely you don't know the secret!" She stood dumbfounded, sure he was mistaken.

"I confess freely that I do not know everything... I was never fully trained!" A clue which might not be so obvious!

"Platinum is a complex element to dissolve! It needs a much higher temperature! That's why you must use, instead of wood..."

"Bone!" Samson interrupted. "Animal bones which have been scraped and dried." Aleesa frowned, disconcerted that this stranger knew the very secret which she had hidden to secure her husband's continued honor.

Samson finally saw through her mask. And he remorsed for her. His own absence over the years had depleted her faith in her son's continued existence. To her The Legend of Samson Trellis had become irrelevant. The very promise which Simon Sr. had made to the entire nation had given faith to everyone... except the woman that he loved the most of all!

"You reveal more than I expected!" she continued. "The secret of bone, I had thought, had died with the expiring of my own husband's breath!" Aleesa was ranting in rage. As far as she was concerned this complete stranger had depleted her husband's right to honor! She asked one last question.

"With whom did you apprentice?"

That last question hit Samson to the core. He could no longer hold back his tears of grief. He remembered the love his mother had always offered... but he now felt that the mask over his mother's eyes was irrevocable.

His father, the king, had been slaughtered on his own land! At his own home! He himself had seen his father fall. Andrew Trellis' death, Samson believed, had been a clear cut case of murder. High treason! Treason which had led to twenty years of civil war within their own homeland. A homeland which he himself had always though would some day be his own to rule!

Samson experienced a physical breakdown, sobbing his pain for his dead mother who was truly dead to him now! Aleesa saw the man fall to his knees and knew that this conversation had distressed both of them.

Her stubborn mood melted in obvious sight of a very disturbed fellow dignitary who needed her guidance. Being naturally a loving, kind woman who was usually very sentimental, she lowered herself to his side and, remembering her dead son, slid his head to her knees. She did understand pain... they had both suffered from this incident.

Something glistened in the failing light... she looked down... she saw the outline of a sheathed knife. When she looked closer she noticed that it was a dagger made of pure platinum! Whoever this man was he was indeed talented.

The man seemed to be recovering somewhat from his apparent distress. Samson noticed the dagger that his mother held to the side. He had one last possibility... and if she remained blind to it... then his mother's spirit was indeed dead! He reached for the dagger and raised the handle to Aleesa's eyes... he was deliberately forcing his royal crest into her sight.

"You don't even recognize your own son," he grieved in anger. "But then again, look what being separated for twenty years has done to us!"

Aleesa saw the Eagle on the dagger... she had seen and even held... used this knife a million times before! She considered it to be the greatest of her own husband's creations! And a treasure she had been sure was lost forever! Samson had kept it all this time! The return of the treasure from her own beloved husband cleared her eyes forever.

Aleesa's eyes had been watering for five minutes... she had much trouble accepting this most miraculous truth! Her mouth finally sagged... then she screamed her tears of clarity! The dignitary who had come to visit her was indeed Samson Trellis... she suddenly saw how blind her masked feelings had made her. She reached out her arms... it was now Samson's turn to comfort the newly resurrected spirit of his dead mother!

Samson knew that she had not cried often but his absence had led to doubt and regret which had never been released.

Aleesa sobbed in her eldest son's arms and finally saw how stupidly blind she had been... and she finally realized that The Legend was based on faith... which she now realized had been lost to her for eons.

"My poor mother..." Samson grieved as he combed her locks with a hand. "How hard a time I had resurrecting you!"

Coronation and Acknowledgements

The door to the office finally opened. Somehow Simon knew what would come about. Aleesa and Samson entered arm in arm. The queen had her head on Samson's shoulder. She cried fresh tears. Satira moved toward her... but Simon held him back. He knew that this was the moment the whole nation had been hoping for.

Aleesa lifted her head off the shoulder. Then, suddenly, Samson was prostrated at her feet. She cried all the harder. She knelt beside him and put both hands on Samson's head. She knew what Samson was asking and her heart's greatest need came out. She breathed heavily as she knew this must be done in public view.

"As queen of the Dwarven Territories I hereby acknowledge my right-born heir! Samson Trellis!" She sobbed the name. Then Samson rose. He did a royal bow to the queen. Then he comforted her once more.

Satira cried with surprise. Simon was the first one kneeling at the royals' feet. All others followed. The coronation was complete. Samson Trellis had finally been acknowledged.

The party rose and took their seats. Aleesa cleared her eyes and stared at her long-lost son.

"It's time for your coronation speech." she revealed. "I have officially named you my heir. As such I have now relented my royal power to my son! Therefore you are now official Royal King of all the territories!" Samson felt stunned. He knew that the royal crowning would follow. He reached for breath and spoke.

"As king..." he sighed with confidence. ""As king my first acclamation is the choosing of a royal bride." Aleesa jumped with surprise. Simon cried his pride. He knew who Samson would choose. "I choose the daughter of Alanah Milner, Myna Milner, honorable healer of the Dwarven Territories."

Aleesa's eyes widened. Then she laughed her surprise.

" I await the acceptance of Myna Milner as new Queen of the Dwarven Territories." Aleesa cleared her tears. But she fell back into deep laughter. Satira approached. He was near tears of joy too!

"What's so funny?" Samson asked.

"With Myna Milner as your wife Alanah will become Queen Mother!"

"I know that!" Samson responded.

"It's really ironic..." Aleesa held her hand up to speak. Satira relented.

"After my attack at the Trellis raid, Alanah Milner was the very healer who nursed me back to health!"

"We had never found a way to repay her for saving the queen's life! And now... she becomes the Queen Mother! That is her god-given right! And god bless her for it!"

Samson stood stunned. How on earth had he managed to choose the daughter of the woman who had saved the Queen's life? Of all the settlements along the border, how...? Samson gasped in stunned fascination.

"Samson, I heartily approve of Myna Milner as your chosen bride!" Aleesa acknowledged. "How on earth could this have come about? Son, I think it was destined! Somehow someone out there knew that she would be honored in due time." Samson still hadn't recovered. Then Aleesa spoke once more.

"Go to your bride, Samson. Be with her! Join with her if you please! She's now new queen of all the nation." Samson strode toward the door.

"Cousin!" Simon called. "You go to your queen, I expect!"

"Indeed... I do."

"Then take her! And reproduce! Bring a young prince to us to follow after you!"

"I know that this coronation is not yet totally official!" Samson accepted. "My application as king must yet be brought before the council. But as heir apparent I hereby go to my chosen bride! And whether we reproduce will be decided by both of us! I take my leave now!"

"Good day, your highness!" Simon offered. Samson walked out the door.

Spilling His Past

Samson stormed into Alanah's home. He caught Myna's eye. He was smiling his glory. Alanah went to move to another room. But Samson called to her.

"Please, my lady, don't go!" Alanah turned to face him. She held tears in her eyes. Samson grabbed her hands hard in his own. Then he spoke from his heart.

"I owe you a royal debt of gratitude!" he revealed. Alanah wasn't sure that she understood. "Your gift saved the queen's life!" Alanah gasped. "You were the very healer that nurtured my mother back to health." Alanah turned away in frustration.

"It's a secret I never revealed! I was just doing my duty as a trained healer!"

"And for that I give my royal blessing! Therefore, I reveal your new appointment as Queen Mother!" Alanah shook her apparent refusal.

"How can I accept? I was only doing my job!"

"But the patient you saved was the queen herself! She feels that she has never found an adequate way to repay you! But to be Queen Mother your daughter must first accept my nomination as Queen Apparent!"

Myna stood stunned. She did not think she quite understood. Samson turned to her.

"Myna, I have been coronated King Apparent by my mother! And I chose you as my bride!" Myna turned her back. Samson moved closer and held out a hand. "Please, Myna, join with me... as my equal!" Myna still stood with her back to him. Then she suddenly lifted both arms high and brought them down with sudden force while screaming at the very top of her lungs.

"Is that a yes?" Samson asked softly. Myna immediately turned to him and offered her hand. Samson bent over and right-honorably kissed her fingers. Alanah stood to the side... her secret revealed!

"If the council accepts our nominations then the matrimony can follow!" Alanah stood away. Myna turned to her mother and held her arms out in beseeching quest.

"Mother?" she begged. Alanah ran right into her daughter's arms. She cried with pride.

"You are so blessed!" Myna hugged her long and hard. Then she turned back to Samson. He looked at her with considerate love. Suddenly the look became intense. Myna sensed the sudden lust that was building in her mate. His eyes showed his every need.

"Mother, could you take the boys bare back riding? Give us..." Myna's eyes had not lifted from Samson's. "two hours." Alanah understood their intentions. She bowed greeting her.

"Of course, Your Majesty." she offered. Myna gloried in her new title.

"No..." Samson countered, holding his hand to her. "don't go yet! We need to talk!" he told his queen.

"About us?" Samson nodded.

"You may leave us... but only to another room! My bride and I must have personal privacy to talk!" Alanah bowed low and went elsewhere. Samson sat and called Myna to join him.

"We have a pertinant mission to accomplish... a duty to all the nation!"

"Bring ourselves an offspring!" she acknowledged.

"As soon as naturally possible."

"Then why are we delaying?"

"Because... there is much you don't know!"

"About... your past!" Myna hypothesized.

"Yes. I am about to reveal the truth of where I've been and who I was!"

"I will listen attentively!" Myna promised.

"When I was forced out of my homeland Simon's men took the duty to guide me to the border. That particular boundary was the very place I had come from before we met. I traveled alone from then on. I knew that the most prominent and safest direction would be to the west!"

"Because of the river!"

"Very insightful! You know your geography well. Before I reached the river I encountered a reddish furred beast! A hellhound! " Myna squirmed with fear.

"It was my first encounter... but wouldn't be the only! He literally jumped at me. The struggle was long and dangerous! The animal singed my hair with its own breath." He spoke quietly. "I knew that I had little time left. So I ended it!"

Samson took out his dagger which his mother so adored.

"This weapon is my life source! It's the only weapon I ever used! It has taken the lives of ten of them!"

"I started to leave... but then a thought came to me! For my own survival I had to learn as much about these beasts as I could. So I gutted it. The ribs had been showing. It had been suffering from starvation. Now I knew the reason for the attack. I noticed that the animal was indeed male."

"I learned the positioning of each of the organs... and now I knew which area to attack! Next time I would try to pierce the heart! I also learned how the creatures manage to produce fire."

"On the day that I reached the river I praised with glory! For on the other side stood civilization."

"The town itself is called Colina... it's a waterfront city. When I first arrived I stayed at an inn... Boar's Head Inn." Samson paused remembering the bar he had just left recently.

"I soon learned that it would be my regular pit stop. I bought a house close to the town wall. Despite its locale the town is surrounded by twenty foot walls. There are two gates which are constantly guarded."

"The years passed. I got to know the barkeep. We shared good times talking. He was my confidante. I knew I could trust him. His favorite verse was, 'The Barkeep Is A Drunk's Best Friend!'"

"I worked as a travel guide... I knew that becoming a smithy would make me stick out."

"There came a time when Leo Anasthasius had proposed. But his fiance, we later learned, stood barren. I escorted her royal majesty from Anasthasia to Colina. When I arrived back I encountered a neutral cleric who offered to take me in as a member of his band. I jumped at the chance!"

"We became the Avengers... then, three years later, there were eight of us! So we renamed ourselves The Eight Avengers! We conquered two missions in a five year span."

"The first was a mission concerning a chaotic magic user who had captured Ludwig Wildlife, the chief headman of the Elven Territories. The person who asked for our help was called Sasselia! She was Ludwig's daughter. She holds the title of Princess of the Elven Territories."

"We infiltrated the shelter. But he escaped, leaving the princess with burn marks on her hands."

"What happened?" Myna asked.

"Andrew left on horseback... but Sassy somehow caught up on foot! She grabbed for the reigns... Andrew drew a dagger... so she lowered her hands to the side reigns and began to drag. The horse was going full trot! The reigns began to cut into her flesh! The flow of blood made her lose her grip. She landed in the dirt crying for vengeance. We later caught up to Andrew and Sassy finally got her revenge."

"We later learned that the second mission was directly connected to the first."

"The mission was instigated by the kidnapping of the Elven Princess from a matrimonial service... the culprit was her own half sister! This mission came to be known as The Wildlife Campaign."

"Sam, Sassy's half sister, brought Sassy to her hide out deep in the Black Mountains. So we headed to the territories once more. We later learned that Sam was mother to Andrew's six year old son!" Myna opened her mouth in amazement.

"This was more a family feud than a campaign. But we spent much longer on this mission. The spiritual aspects of this mission were intense. We went around the Black Mountains to Galala Valley, an unknown land beyond the mountains which, for generations, the Elves had been sworn to protect."

"Why? What's so important about that piece of land?" Myna wondered.

"Because the valley holds a secret which the Elves have been sworn to protect! The life form living there is known as Palapany... she is a Centaur. We met with her... but she only comes when she's needed. She takes two forms... when she travels she takes on the form of a cameo colored horse. But her Centaur form is.."

"Horse and humanoid!" Myna finished.

"You know your fables well! The elves revealed that their legends are based on fact... past experiences which have been thoroughly documented."

"We secured the territories once more! But we spent much time on this campaign. Suchi was grief stricken with his loss. He soon decided that our troop had to be disbanded! The territories family was deeply involved in Sassy's rescue!"

"After her safety was secured she married her fiance of ten years... it was an arranged marriage. She now has a two year old daughter. Anna Littleton!"

"When we arrived home people became concerned. Suchi discussed the disbanding with Baron Brunweger. We decided to go our separate ways."

"We had already lost Sassy to matrimony... and a pertinent couple who had chosen to settle in the Territories. We were breaking apart!"

"Then came the day of Leo's conference officiating the cancellation. The Eight Avengers were all present. The council chambers were wide and spacious. Leo Anasthasius himself stepped in. The band began to bow to his honor. He flicked his refusal."

"I will not accept your bow. For I am your equal!" he said. We all gawked. He offered eight chairs in line. He himself sat facing us! Then he spoke."

"Due to the dangers involved in the policing of the Principality the band named Avengers is officially disbanded." he said. "I humbly thank you all for your devotion to this force of power. But, you see, the policing was totally unnecessary! I had alternative reasons to unionize this council."

"Sasselia Wildlife, as princess of all the territories held by the Elves, was in deep distress. At that time I learned a secret which we all have always kept!"

Samson cried suddenly, remembering this unspoken truth!

"Sassy has given her time and love to this band, giving up her royal position, if only for a time. And I proudly call her my equal!" Samson smiled his pride at the memory of her doctrination.

"But there is one of you which is truly royal... and that is you, Samson Trellis!"

"I literally gawked with exclamation! I cried tears of glory to this man, even though I did not know how far this would lead!"

"I reveal to you, he continued, my royal prince, that even though you feared for your life, I let you feel it! But I have always secretly protected you! The fear you felt would be the power which would motivate you during the intense training which I have instigated to you. When Suchi approached you, it was not to become one of them, but to train as my own apprentice! Then suddenly the king of all the principality fell to tears."

"Leo Anastasius had trained me as his own apprentice, he knew the risks, he knew my life was indeed still at risk. But he had trained me himself. He made one last comment."

"Samson Trellis, he said, even though your father was killed in an act of treachery I was also your father! Your master! Your leader! As far as I am concerned I think of you, prince Samson, as the son I never had!" Leo sobbed the last words, barely able to finish his own sentence."

"Peter Huntingham was my contact... the contact that Leo himself had placed there to keep me advised of the war's progress. And now, with the seize fire, I would return to claim my throne."

"I cried for my own fer. I also cried because, despite my loss, I was still loved. Through the power of one man... Leo." Samson cried, swallowing back his love, and continued.

"Sassy immediately rose and hugged me! She was the first at my side! And we both cried!" Samson sobbed quietly, remembering all the glory and love that his fellows had offered. He had told all of them... they all knew that he had been born of royalty. And he had also learned that even Leo knew. His life had been secured in the loving hands of his new-found father.

Myna rose and took his hand. She bent to kneel taking him down with her. She offered him a kiss on the cheek. But he silently sobbed, ignoring her. She observed him closely. She saw love... for all his friends! His eyes became glossy with held back tears.

"Let it out!" Myna whispered. "Let it all out!"

"I'm through crying! My past is gone! And now, if I can, I will bring my nation to glory like never before!" He turned to her and made to kiss her. But she flecked her head back.

"What does Sassy look like?"

"Why?"

"Because I think she was the one emotionally closest to you."

"Sassy!" He slid the name between his teeth. "Blonde colored hair, small pointed ears, thin lips with a pudgy nose."

Myna kneeled for ages soaking in this information. It sounded like Samson was describing Myna herself! Then she realized something else!

"You loved her! She was your first love!" Samson struggled with this truth. But he suddenly admitted it.

"I loved her! She was my blossom! To her I was another dad, even though we were of similar age! We spent hours and hours together! We'd spend nights sharing our own thoughts and open sky as mates would. The only difference was... it never happened!" Myna looked deep into his eyes, knowing what he meant.

"The act!" she asserted. Samson smiled despite himself and nodded.

"The act! Intercourse itself!"

"You were pained... but please don't give up your past!" she turned his face toward her. "Samson, would you look at me, please? Now, look me in the eyes. Color my eyes hazel green."

Samson looked. Suddenly Sassy's image was in his vision. It was like a virtual backflash!

"Sassy!" Samson stared into her face. The pudgy nose was slightly smaller but the same. Her lips were thin but the smile did not spread as wide. The ears, he noticed, were slightly bigger! But besides those differences, and, of course her eyes, Myna Milner was a physical replica of Sasselia Wildlife.

Dwarven-Elven Pledge of Allegiance Samson Leader Lives!

Samson continued to stare at Myna. She looked him straight back! Then Samson began to touch.

"Is it similar?" Myna asked.

"It's remarkable! Except for size there are many similarities!" Myna suddenly broke out in tears.

"I've finally found her!" Myna exclaimed.

"Found who?"

"My soulmate!!"

"What?" Myna grabbed Samson's hands in her own. Then she made a confession.

"I've always wanted a sister!!"

"And Sassy would be like that!" Samson realized.

"Three sons and only one daughter! Talk about tough!"

"So Sassy is the sister you never had!"

"Exactly! My soulmate! Why else would you feel so comfortable with me?"

"I do, don't I? It's like we met before we really did!"

"Sassy gave you my image. But I'm the one who can give you her love!" Myna drew close and rubbed noses. Samson kissed her back. Then she realized something, jerking her head in realization. "Wait here!" She then went into her bedroom. She came back holding a piece of parchment. She handed it over.

"Dwarven-Elven Pledge of Allegiance! What's this about?"

"Read on!"

"We, as the royals of the Elven and Dwarven Territories, hereby acknowledge our dedication and devotion to each nation in recognition of our own similar genetic ancestry. We are of one nation in two territories. We hereby acknowledge that Elves and Dwarves are distant relatives and members of the same people! Therefore, we sign this Pledge of Allegiance between our peoples! Signed, Horticus Trellis Isidius Wildlife

Samson looked up at Myna. "You know what this means?" he asked. Myna nodded her head.

"Dwarves and Elves have the same ancestors!"

"But the theory is they're not sexually compatible."

"I think any woman can be compatible with any man!"

"If Elven and human works..."

"Look at all the mutants that exist!"

"Then the theory of incompatibility..."

"It's a stigma! Something spread to keep us apart!"

"But we're joined at the hip! By my grandfather and Sassy's."

"Look at the date!"

"January 1, 1326! Only fifty years ago!"

"And, I think, this pledge is still active!"

"That's questionable! Both these men are dead now!"

"Then we'll resanctify it! Bring it to Village Wildlife! I'm sure Sassy would agree."

"That's something that could be very difficult! The councils from both nations would have to be advised!"

"Let's think on it!" Myna suggested.

"Now I know why Sassy's your image. You share the same genetic traits."

"Except I'm short and she's tall!"

"That's no big deal! Size doesn't really matter!" Alanah suddenly passed through the room.

"Hope I'm not intruding! I'm just going to take a trip! There is much to arrange for the matrimony! I'm going directly to see Aleesa! And I'm taking the boys!"

Samson and Myna both smiled their knowledge. They would be totally alone!

Myna and Samson were walking in the back yard. She had insisted that they go for a walk. Alanah and the boys had already left. Myna sat on the grass and began to remove her leather shirt.

"What are you doing?" Samson asked.

"Sassy never shared the sky properly so I'm making up for it!" She soon stood topless. Samson stared. Her breasts were smallish but ample. The nipples jutted in the warm air. Then she removed her bleachers. She stood bare except for her personal coverings. Samson prepared quickly then lay down beside her. He nuzzled her ear. Myna lay in the fresh air exposed. Samson gently bit a nipple. Then he moved down kissing her belly button. Then lower still until he undid the clasp of her covering. He stood silent and breathless at her beauty.

Laying down he nuzzled close to offer his warmth. He could not wait any longer. He mounted... kissed her full and put fingers at her clit. Water came... but he still knew it would hurt!

"The pain's only temporary!" Myna reminded. Samson aimed his tool and began the penetration. Myna gasped her discomfort. Samson smoothed his fingers across her entrance. More water came. But the pain was still there. He was half length in. Then he lost control, plunging deeply. Myna yelled out in pain with her eyes forced shut.

Samson was totally buried. He lifted and lowered. Then he began faster. His very form shook with need. Myna opened her eyes and put a hand behind his neck, pulling him close. She then kissed him.

The pressure increased as the human pump began to suck. Samson shook with excitement. Two minutes later he came! Myna gasped her surprise and smiled.

"Babies on the way!" she exclaimed.

Samson laughed his satisfaction. He kept his tool in her until it finally deflated. He then rolled off.

"Next time, let's do it from behind!" Myna suggested.

"I like to hold your wrappers." he confessed.

"M y Samson Leader."

"My Sassy." Samson returned.

"Samson Leader is the man I want to marry! Someone who knows the land and can travel with me. I've never been out of the Territories!"

"I can fix that!" Samson began to rise once more. Myna smiled her knowledge as she observed his stimulation. Myna turned over and separated her legs. Samson mounted from the back. His tool immediately entered by incident. He noticed that sex from the back made the tunnel much more accessible. His knob was hard but he was also beginning to feel sore.

Myna suddenly closed her legs together. Samson put his legs on either side. The pressure from the closure made the tool much more intensely pressurized. He rose and fell. Then he continued. Raising his chest with his arms he put all his back pressure into the endeavor. Soon after he climaxed once more. He rolled off feeling exhausted and sore. Myna rolled onto her back and placed a hand on Samson's hairy chest.

"You think it'll rise?" Samson asked, looking at his deflated and abused tool.

"Maybe later tonight!" Myna suggested. She got dressed and sat down as Samson caught his breath. Then he got dressed and they both moved to the house.

"I have a question."

"Shoot!"

"Who's going to be king? Samson Trellis or Samson Leader?" Samson thought on that.

"I think Samson Leader would make the better king."

"That's what I wanted to hear! In some ways Samson Trellis died ages ago!"

"Confirmed." They finally entered the house and sat on the living room couch.

"Will you swear a vow on your throne?" Myna asked.

"Depends what the vow concerns."

"Some day, when you get a chance, I want you to tell Sassy how you feel about her."

"Why?" Samson seemed confused.

"Because things change... she has been, is and always will be your first love! When I die I want her to have you next!"

"You're willing to share?"

"She is my spiritual sister and soulmate."

"Nothing's about to change any time soon."

"Besides, I'd like to visit Galala Valley... some day!"

"Then it's a date!"

Secretive Meetings

Later that same day the Waterbee Trio, followed by Samson and Myna riding side by side, set out for Waterbee Settlement. When they arrived the sun stood below the horizon as dusk had also arrived. Simon Sr., or what Samson took to be Simon Sr., stood in the courtyard and observed them dismount. Simon Jr. stood down and began introductions.

"Father, I would like to humbly introduce Samson Trellis, King Apparent, and his fiance, Her Majesty, Myna Milner." Samson and his bride shook their greetings. Simon Sr. took the liberty to kiss the Queen Apparent's hand. Then everyone strode toward the house as a servant took charge of the horses.

"It's been a long time! We have much to discuss! I have heard of your coronation, and tomorrow, the council has been assigned, and will convene to approve your nomination. The crowning will, of course, follow!" Simon Sr. sat in the living room. Myna and Samson chose to sit across from him. Samson looked down and noticed that Myna had started holding her tummy! He gently put one of his own hands over hers. Simon Sr. obviously noticed.

"Expecting?" he asked.

"We certainly hope so!" Myna punned. Simon silently nodded his assent.

"It would indeed be an asset!"

"Simon, I want to thank you for all you have done for me!" Simon Sr. just waved away his gratitude.

"The legend has indeed come true!"

"A legend that you yourself instigated, I may add!"

"I always knew you would survive!"

"I myself had many times when I felt I may not! But now I understand!"

"Anasthasius freely took you in."

"It had been arranged, then."

"After I rescued you I immediately sent a scribe to Leo asking him to act as your guardian! After all, a twenty-two year old smithy apprentice cannot lead a whole nation by himself!"

"I do realize, now, that at the time I was a very ignorant prince."

"And yet you were still willing to stand in front of your mother standing ground against an enemy! You do realize that your own life is paramount when compared to your mother's!" Samson freely nodded.

"Alanah Milner was the healer who nursed my mother back to health."

"Yes, but even so, why this youngster? Of all the beauties in the nation, why did you choose Myna Milner?"

"Actually, I will freely admit, she was the one who chose me!" Simon stood stunned but soon caught himself.

"Why, my dear?" Simon asked of her. "Why Samson?"

"Samson is well traveled." Myna responded. "Older and truly committed. As far as I have always been concerned, my mate's personal titles would not be a consideration for me! I honestly felt that if I didn't clasp my grasp on this one there would never be another as adequate!"

"What is it about older men that appeals to you?" Simon asked next. "What is it that really turns you on?"

"Mental maturity! My mate must be at my own level! Samson is a well-known, considerate man! He will indeed make a superior leader!"

That night Samson and Myna were assigned a room at Waterbee Settlement. Myna was soon undressed and under the covers. Samson, however, took his time. When he did, finally, join her he smiled with knowledge.

"It's later!" Myna joked. Samson caressed her tummy.

"You think one's started?" She shrugged her shoulders.

"I'll know for sure in a couple of weeks!" Samson nodded his coherence. He gently kissed each of her nipples before mounting again. Samson swore that this would be a long, joyous session. He made considerate love with her for two whole hours until Myna suddenly noticed that he had fallen into deep sleep. Myna smiled, as she loved to feel his gentle weight on her. She also fell asleep, her hands behind her head, Samson's weight a comfort as his head rested on her shoulder.

The Dwarven Council was convened. In attendance were all the old war's surviving generals as well as Aleesa and her royal son. Aleesa officially opened the proceedings.

"I proudly introduce my son... Samson Trellis! I have freely coronated him King Apparent! I now ask for your consent to the crowning!" General Roger Sentium spoke up.

"I need clarification of this man's training and ability to rule with an iron fist!

"I do realize," Samson spoke up, "that you folks know very little about my continued training... but I assure you all, I am skilled!"

"I humbly dispense this document," Simon spoke up, "directly from the office of Leo Anasthasius who has been teaching our King Apparent as his apprentice. It details Samson's continued training in military combat spanning the whole twenty years that he was gone!"

Roger spent the longest time of all the councilors looking over the parchment with Leo's business seal.

"This is very impressive! Cross bow, dagger throw, large and short sword battle, archery as noted from previous training, of course, as well as lancing. Large sledge attacks, secret door training, lockpicking, grapple hook knowledge and adequate war horse experience! Very impressive indeed! Unless there are any further questions or objections I humbly vote for, and confirm, Aleesa's wise actions. Samson, I personally knew your father, Andrew Trellis, very well! He was one of the best leaders this council ever had. Of course, he was a descendant of the Trellis line! And I'm sure you will be ten times the leader Andrew was. Your own military training surpasses any knowledge the Territories ever had!"

"He has chosen a queen?" someone noted to ask.

"A queen has already been chosen and nominated on my son's behalf! Myna Milner of Milner Settlement has been nominated as queen, and Alanah as Queen Mother." Several around the table gasped in surprise. "We all know how Alanah has contributed, personally, to the medical side of this war!"

"If it weren't for her you would be dead!"

"Or barren!" Aleesa put in.

"What?" was heard all around the room. Samson looked on, his face turning white.

"It's the truth!" Aleesa confessed. "Alanah said that the puncture was imbedded in abdominal bone... but the point stood very close to my womb! She said that the head could penetrate or pierce it if I moved around too much! The trip by horse had caused some damage but hadn't quite killed my fertility... thank goodness! Which brings me to the next subject... I expect that Myna may already be in waiting!"

"Well... my word!"

"That indeed was fast!"

"Let's vote on the nominations... do we have a unanimous decision?" All members agreed. "Then the crowning and matrimony will proceed. Schedule it for tomorrow." The council began to scatter. Samson took his mother's arm and led her out. He swore he would ask for more detail about the fertility.

Samson and Aleesa arrived back at Anders Settlement. Neither had spoken about the fertility on the way. Samson took a seat and called her next to him.

"What were the real stats?" he asked.

"The wound had caused a 5% chance of barrenness."

"5%. Please continue."

"Alanah told me in private that I might be barren. I kept that secret to myself at first! I spent two months in bed recuperating! I realized that I would not be returning to Trellis Settlement. The fires were burning at the time so Satira offered to let me stay as long as I needed. I procrastinated, busy with war tactics and the future of our nation."

"Satira was one of the generals throughout the war and it surprised me how he always, somehow, survived front after front! My respect for him blossomed. On the 2 year anniversary of my stay at Anders Settlement, when Satira had lost twenty of his best men, he made sure to be home! He told what had happened, but afterward he proposed to me! Asked me for my promise! I was dumbfounded... I didn't know what to say! I desperately wanted to agree but something still stood between us!"

"The secret!" Samson noted.

"I took the liberty to ask if he planned for a family. He said it was only one of his plans for us! I knew I could no longer keep it to myself! So I told him! Then I asked if he still wanted me. He told me that this would not be an arranged marriage! I had lived with him for two years! All that time he had observed and listened. He had learned to love his compatriot in arms!"

"You agreed!" Aleesa nodded vigorously.

"That very night was the first inkling of sexual activity I had enjoyed since the raid. He told me he had faith in my continued fertility! Two weeks later I missed my period! I was ecstatic! I told him that night... he said, "I told you so!" but I didn't mind!"

"So you consider your young uns a work of magic!" Samson offered. "I would love to meet them!" Aleesa freely agreed. She called her children, Samantha and Peter Anders.

Peter came in running... Sam crawled in slowly. They both suddenly laid eyes on Samson! They prostrated themselves at his feet.

"My young siblings, I greet you both!" Sam looked up to her elder brother. Her hair was long, blonde and styled very adequately. Samson noticed that she took great pride in her own appearance. Then he looked at the younger sibling, Peter.

Black hair the color of Satira's, a long nose and thick, well rounded lips hid an inner power... Samson knew that this boy would be an excellent soldier! His dark skin made for the appearance of invincibility! He was, indeed, Satira's son!

"My liege, I would speak!" Sam notioned.

"Then continue."

"We have always held the legend blessed! We had much faith in the return of Samson Trellis! I personally tried, hundreds of times, to spurn some essence of faith in our mother! But the light of her faith was gone!"

"But now it has rekindled in the true return of Samson Trellis! Stand and face me, both of you! I will make a proclamation.!"

"Samantha, as the elder of the two, you stand as my man in arms and head general of my troops. Peter, you will be my head general in charge of troop movements! You will both be given power within the Dwarven Council and stand as my humble lieges! Active immediately!"

Aleesa looked at Samson, with some surprise, and cried tears of joy! Her family was, indeed, whole once more!

The Legendary Samson Trellis To Wed

Samson Trellis Indeed Lives!

The return of Samson Trellis to the Dwarven Territories has led to the end of 20 years of brutal war! And now he is to be crowned and take his place on the Dwarven throne. Along side him will sit Myna Milner, new queen and icon for the nation! It is also said that Myna, honored healer of the Dwarven Territories, is expecting their first child! Hopefully we will be blessed with a young prince!

Samson says that as king he will be known under the title of Samson Leader! This is the name he has used for the past twenty years working as a travel guide throughout these continents. He will, indeed, make a superior King!!

Peter Huntingham read the front page of his copy of The Daily Scribe, the town paper for Colina. He smiled his knowledge... he definitely had to send this article to Leo!

Amin Nanderson also read the same article further back in The Weekly Writing, his local and distant Political Paper! He literally broke out in tears of sentiment!

Crowning, Matrimony and a Revamp

The crowning and matrimony had finally arrived! Samson stood on a platform facing a Dwarven Sage, with his chosen mate standing next to him. The Dwarven Hall was crowded beyond limits! People stood outside the building hoping to touch the royal couple! The sage opened the proceedings.

"Dearly beloved, we stand here to meld the King and Queen in holy matrimony! If any have objections to this joining you must speak now or forever hold your peace!"

The room stood in complete silence! But suddenly two personages began pushing their way forward!

"Hold the ceremony!" one called. Samson recognized that voice! He turned and glanced at the intruders. He saw the forms of Suchi Evil Killer and Lord Eugene Brunweger. Samson waved his hands to let them through as he dismounted the stage.

"Hello, your royal majesty!" Suchi greeted. "We thought we would crash this great celebration... the greatest in all history!" Lord Brunweger shook Samson's hand vigorously.

"We will talk after the proceedings!" Eugene suggested. Samson returned to his bride. The ceremony then resumed.

"Samson Trellis, do you take this woman, Myna Milner as your lawfully wedded wife? Will you recognize her equality and respect her opinions as long as you both shall live?"

"I will!"

"Myna Milner, will you take this man, Samson Trellis, as your lawfully wedded husband, to lead the Territories as his equal while respecting his power over you?"

"I will!"

"Then the crowning can follow!" the sage announced. Both Aleesa and Simon Sr. approached. Two cases were brought before them and were opened. Both crowns were made of pure platinum with diamond emblazons on them! The queen's crown was placed first. Then Samson's King crown, the one his own father had worn at many a ceremony, was placed.

"I hereby pronounce Samson and Myna Trellis King and Queen of the Dwarven Territories! You may now kiss the bride!" Samson took Myna in both arms and kissed his new wife long and hard! The crown suddenly stormed their joy! The royals spent the next fifteen minutes among the crowd and then gently began their leave.

Suchi and Lord Brunweger were honored guests at a private supper that Aleesa had prepared the day before. News had spread about the crowning and the two leaders of the old Avengers troop wanted to be here, naturally! Sasselia Wildlife was living at Village Littleton, Suchi passed on, with her husband Alvin, who had retired from the Elven Guard. They had a two year old daughter named Anna.

Thonolan and Tralina now had a two year old son, Thor, who was turning out to be a natural talent at sword play. Thonolan would take Thor on at sword sticks and more often than not Thor won out! "But then again, what does a neutral thief know about sword play?" Suchi asked.

Roger and Roderick Bloodbath were currently gate guards at Colina. And they were climbing the ladder very quickly within those ranks, it seemed.

Later that night, when the guests finally departed, Samson had a long talk with his mother. He would revamp the Dwarven Council... he had been thinking long and hard on how he would change things. But now he had made his decision. Aleesa convened the Dwarven Council, under her rule, for the very last time!

The council convened the next day after all had been arranged for an emergency meeting! The generals all took their seats. Aleesa and Samson remained standing. Then Samantha and Peter Anders both appeared. Samson placed them to the side, standing, right next to their mother. Aleesa addressed the council.

"As you all know, I am officially invalid as to the leadership of this council! I now relent my post to my son, King Samson."

Samson stepped up to the table as Aleesa returned to her spot. He then leaned on the table and spoke.

"As king, I have decided to revamp this council to include new blood! Several of you are aged... I mean no disrespect, but if this council is to continue under my leadership it needs young blood with original ideas. I hereby ask three of you, those who are over the age of seventy-five, to relent your seats to them!" He waved a hand toward the standing youth. "The retirees will be; Simon Sr, aged at 82, Roger Sentium, aged at 79, and Arthur Hillington, aged at 92! Please make way for these new members!"

The named councilors freely relented their seats and stood together in the opposite corner. Samson placed Samantha in Simon's seat, Peter in Roger's seat, and he gestured to Aleesa to take Arthur's seat! At first she refused.

"I'm not qualified!" she argued.

"I'll decide who's qualified!" Samson responded.

"Now, the council is fresh, and we will proceed." Satira motioned.

"I would be willing to relent my seat to someone!" he informed. Samson shook the negative.

"I need you!" he insisted. "This new council will rule, unchanged, under contract, for the next ten years! There will be no changes, except through retirement or demise! This council's retirement age will be 75! I want my mother, Aleesa, on this council until her 75th year. Or until the time of her demise, whichever comes first!"

"Satira, you, along with your son, will be my head generals! Samantha, I have promised, will be my man in arms! Directly ruling under me! Satira, you will take Peter in hand, and lead him well! Starting now we will start basic training for each of my generals! The same training I received from Leo will be passed on to each of you!"

"Some notes: during the proceedings of the war, my mother has informed, there was much argument and insubordination among you! Under my rule, there will be no argument... no indiscretion! Each councilor will be given a chance to voice himself. Speaking out of turn will be denounced. There will be no argument. What happens in here affects the lives out there! You will be recognized in good order!"

"I humbly thank the retirees for their prolonged service during these past twenty years! But now, with fresh faces and a revamp of policy as well as members, we will be a council to reckon with! That is all!"

A Business Trip to Village Littleton

Seven months later Myna began her labor. She was only eight months gone, but, Alanah knew, that was not unusual. Twenty hours passed during which time Samson relaxed in waiting of his own with Simon Sr. and Satira both in attendance. Alanah suddenly came out of the chamber at Anders Settlement holding an infant in cloths. Samson immediately went to her and looked.

"You have your newborn Prince!" Alanah congratulated. Samson went in to see Myna. They spent much time talking. They had both decided that, if a boy, they would name him Margo!

The council prospered under Samson's wise leadership. Basic training was gradually completed for all councilors, and six years passed with the good life lived well.

Samson had rebuilt on Trellis Settlement and now spent most days smithing with his father's salvaged equipment. His first born son, Margo, standing at six years of age, was a constant delight!

Samson and Myna both sat in their living room discussing that day's achievements. Margo suddenly ran across the room. Samson quested him.

"What's your rush?"

"I've got diarrhea!" Margo responded as he exited toward the restroom. Samson laughed. Myna pushed Samson's shoulder in false consternation.

"Don't laugh so! He's telling the truth!"

"I know! But I couldn't help myself!"

"I think it's time we went on a vacation!" Myna revealed. "The nation has been so peaceful these past six years! I'd love to visit Galala Valley! Remember your promise!"

"Okay, then I'll arrange for the trip! You're on!"

"Finally!" Myna resigned. Samson laughed his consent.

Samson had arranged the campaign of troops to travel to the Elven Territories. Their exact destination was Village Littleton, where, word was, Sasselia and Alvin Littleton had settled. The troops traveled out of the Dwarven Territories near Milner Settlement. Then they continued north.

Two days passed as they covered grassland after grassland with very little forest. There was not a single hellhound in sight! And frankly, Samson didn't expect to see any! They usually attacked single persons... never groups!

On the third day of travel they arrived at the small coursing river which they all knew must be crossed!

"I'll go first!" Samson informed. "Then Myna and then the rest of the troops. Wait until I'm across before you come!" He marched into the coursing river! Fifteen minutes passed before Samson reached dry land. Myna followed cautiously. Then the rest stormed the river from several angles. Some minutes later they continued to the north-east.

Several days' travel went by before they came to the Elven Forest from a crude angle. Samson stormed his way through dense brush to find the path.

Samson had entered the forest at an odd angle purposely, hoping to hit Village Littleton from a shorter distance. They had arrived in the middle of the road. Village Wildlife stood far behind!

They passed two neighboring villages before they arrived at Village Littleton in full force. Samson rode to the middle of the village and dismounted. Others followed cue.

The villagers all gawked. But one of them had the courtesy to see to them.

"My greetings, your highness!" she greeted in Elven. "We were not expecting you!"

"I will have word with your sorceress... immediately!" Samson answered.

"Understood!" The Elven maiden led Samson to the Alvin Littleton residence and retreated. Samson caught his breath and then knocked. Sassy came to the door and opened it. Then she literally stared!

She smiled her surprise and laughed. Then she let him in.

"This is some surprise!" she assured. "What brings you to my realm?"

"Business!" Samson revealed.

"What kind of business?" Samson revealed a piece of parchment. It was Myna's copy of the Pledge of Allegiance. Sassy read it. Then she gawked.

"Where... this is top secret! We have kept this pledge one of our highest secrets these past..." Then she looked at the date. "1326? This is a later copy then ours! Ours was signed in 1300!"

"Then you know about it!"

"I always have!" Samson bowed his head in frustration holding his nose. "Samson?" she asked.

"There's something else we must discuss!" he revealed. Then he put both hands to Sassy's cheeks. "Forgive me for this revelation... but my queen has made me vow... Sassy, I have always loved you! As a mate!" Sassy physically blushed beet red. Then she backed away.

"I can sometimes understand that!" she revealed. "We were, after all, best of friends! I spent more time with you than with anyone else! After all, we're both royalty!"

"Then you can share in my sentiment!"

"Indeed I can! But I'm committed! Alvin needs me! He loves me as much as you do! And my little girl..."

"Anna would be eight!"

"Yes indeed!" Sassy began to pace the room. "When did you first realize this emotion for me?"

"When Sam stole you from us! I spent days alone, confused, and lonely for you!"

"I'm sorry, Samson!" Sassy freely admitted.

"I have some compensation!" Samson relented. Then Myna showed up at the door wearing a hooded cloak. The hood hid her face. "I present my queen, Myna Milner!" Myna stepped in and turned to face Sassy lowering her hood.

Sassy looked at her through screnched eyes and physically gasped in shock. She saw the blonde hair, exactly her own length, the pudgy nose and the thin lips. She tried to say something but nothing came. Then she caught her breath.

"How?"

"Now you see my compensation!" Samson smiled.

"She's me! In Dwarven form!" Myna clasped Sassy's hands.

"We were born soul mates!" she revealed.

"And due to the facts contained in the Pledge you two have the same genetic traits."

"My short stature is a dramatic difference however!" Sassy turned to the wall to think.

"What do you want to do with this?" Sassy asked of the piece of parchment.

"We want to reactivate it! With our own signatures! Continue what our fathers began!" Myna confessed.

"Impossible! The stigma..."

"Is non-existent!" Myna countered. "Look how many mutants are born every year! Does that mean they are worse than anyone else?"

"I understand the sentiment!" Sassy relented. "But this is a decision for the whole council of headmen! They must be informed!"

"Then arrange it! We will remain here until the decision's made... one way or the other!"

"Understood! I'll get on it immediately!" She turned to Samson and brushed his cheek in sympathy. "I'm sorry, Samson!" He simply shrugged it off with a smile. Then Sassy left.

"Next stop... Galala Valley!" Myna suggested. Samson nodded in confirmation.

To Galala Valley

Margo Trellis approached from one direction interested in the sights of this very different village. Anna Littleton approached from the exact opposite direction. Margo was not watching his direction, fascinated with the intense height of most of the Elven adults. Anna was walking, head down, feeling sullen boredom. Suddenly both collided in a smash of arms and legs and fell hard on their rumps. However, both just broke out in laughter.

"My name's Anna Littleton!" she acknowledged.

"I'm Margo Trellis!" Then they both rose, Margo helping her up. When they stood they both noticed that they stood at identical height!

"I'm a Dwarf!" Margo offered. Anna gasped her surprise.

"If you're a Dwarf... and I'm an Elf... then why do we look the same?"

"We have similar ancestry!" Margo revealed. "I heard my parents talking about it numerous times!"

"Who are they?"

"Samson and Myna Trellis! The royals of the Dwarven Territories!"

"My mother's the sorceress and head of this village! Her name's Sasselia!" Anna informed with intense pride.

"You don't say!"

"She's also known as Princess of these parts!"

Samson approached Sassy from aside and she turned. "I have a request." he revealed.

"Shoot!"

"I'd like to take my men into Galala Valley! My wife wishes to meet Palapany... that is, if she will even show!"

"I guarantee... she will!" she reassured. "That centaur knows everything before anyone else does!"

"Then we can go?"

"Feel free! This is one decision I can make on my own!"

"I'm hoping to return as soon as we can."

"Relax! Enjoy the sunshine! The council's going to be out, arguing I may add, for ages. This pledge has sure brought this nation some well appreciated attention!"

"What do you think will happen?"

"They'll follow through! That's my thought, anyway!" Samson grabbed Sassy by both arms.

"Just think... our peoples will be able to interbreed."

"Leading to a medium sized offspring! I do understand your motives! At least, the theory is intact!" Suddenly they saw Anna and Margo approaching hand in hand. Sassy and Samson both gawked. Then they locked eyes.

"You thinking what I'm thinking?" Sassy extrapolated.

"Margo and Anna... as a couple!" Sassy nodded.

"Let's just wait and watch! It will take some time!"

"I'm giving you Anna's promise!"

"I give you Margo's!" Samson reciprocated.

Samson and his troops reached The Gate three days later. Margo had been asked to stay at Village Littleton to keep Anna company... which Samson and Sassy both nourished, of course! He had cheered in reply. Samson himself, as the only one who knew the way, led his troops. Soon Samson decided to pass through and carry on.

Samson knew that it would take many days to tour the entire valley so he decided to travel two days in, toward Galala Strait, and then set up permanent camp.

Myna gloried in the beauty of the land! Pure grassland stood as far as the eye could see... and much of the plant life was, indeed, edible! Myna spent most of her time hunting for veggies and herbs for spicing meals.

Three days passed very quickly, it seemed to them. Samson intended to stay in the valley until Palapany did, indeed, show up. A fourth day passed. On the fifth day Myna and Samson sat together, enjoying the clear blue sky as was their custom. Myna suddenly spoke up after a half-day of contemplation.

"How old is Margo now?"

"Six going on seven!"

"Don't you think..." she looked at Samson strangely.

"What?"

"I'm tired of sheathes!" she exclaimed. "Let's get natural!" Samson looked into her eyes and saw need. But for what? Then he connected.

"What you are really saying is you want another child!" Samson concluded.

"Margo is six years old! We should have tried before now!"

"You're probably right!"

"I'm not getting any younger either! I mean, I'm the one who has a time limit, here!"

"You're only 28!"

"But let's just do something about it!" Samson smiled at her confusion and had to laugh internally.

"You got a deal!" he concluded, reaching to shake hands. She offered hers but Samson turned her hand in his and kissed her fingers. They began to flirt. Suddenly Myna thought she heard a soft clap of thunder. She looked up but saw a continued clear blue sky.

"Sounds like rain!" she suggested, feeling somewhat confused. The thunder grew a little louder. But Samson smiled with his knowledge of the truth! Myna stood spellbound. Samson just laid back on his arms and observed the horizon. He knew that Palapany was coming to see them!

The thunder escalated into intense hoof beats! Myna saw, in the distance, a cameo colored horse approaching. She opened her mouth wide in surprise.

The thunder died down as the horse approached within view. Then it walked right up to them. Myna stood up and began to comb its nose. The centaur shimmered and changed form. Before them stood the human horse form of Dwarven and Elven legend. Myna stood dumbfounded at the beauty of this being which she had never seen in person.

"Greetings!" the Centaur spoke with a whispery and soft tone. Very peace-feeling and mellow! "I have come to conglomerate our joining! You have both agreed to continue your journey of creation! By power invested in me I hereby announce that this offspring will be a child of the Galala Valley! I humbly request that you honor me with becoming your child's godmother and I also ask that you include my name in her title! But... first... we must think of the conception! I will take both of you to a secret and sacred place where you will have a world of privacy!"

The troops were all gathered observing their conversation. The children smiled in glee at seeing such a spiritually uplifting image. Samson turned and spoke.

"I'd like to introduce Palapany! She has asked that Myna and I join her for a distance! We will be gone until dawn tomorrow! Carry on!" The troops then turned back to their assigned duties. Samson smiled.

"What now?" Palapany laid flat on the grass, as was her custom, and turned toward her back. Samson consented after some minutary preparation. Then she converted her form back to the cameo horse. The transference took no time at all and they were off at her thundering trot. Her hooves dug into the grass with a thundering jolt. Palapany picked up more speed. It suddenly felt like both passengers might fall off her back! But Samson, sitting behind Myna, protected her and promoted continued balance.

Half an hour passed as Palapany trotted at full speed... however, she never seemed to tire! Myna watched the land go by at a drastic rate! She observed their current location. Samson pointed out the eastern point of Galala Strait but noticed that right then Palapany turned course heading straight east!

She galloped for fifteen more minutes and approached a stone building. Samson immediately recognized it as the Galala Shrine! He laughed his glee as memories from eight years ago flashed through his mind! He knew that he and Myna would have total privacy in the shrine which stood half a day's ride from camp!

Palapany slowed and relieved her passengers. Myna stumbled uncertainly for some time with woozy dizziness, but it soon cleared. Samson pointed to a doorway which stood in a rock wall facing the west!

"This is where it all really began!" Samson informed. "The Wildlife Campaign!" Myna did the honors of opening the door. Samson quickly followed.

The shelter they had entered was fairly spacious but not at all luxurious. The floor was made of dry dust!

Samson removed his pack which he had prepared earlier and opened it revealing two silver chastels with engraved designs. Then he filled each of them from a flask of ale he had hidden. Myna lay down after placing a blanket over the dirt floor and took a full chastel. Then they toasted to health.

"To Palapany!" Samson gestured. They touched silver. Myna took a huge swallow.

"What do we call this infant, anyway?" she asked.

"I think if Palapany has her way this infant will be a girl!"

"I humbly agree! It would be nice if Margo had a younger sister!"

"Then it's decided!"

"What about a name?"

"I've been thinking about that on the way here! Anna Palapany Trellis!"

"Anna... after Anna Littleton?"

"And Palapany, of course!"

"Then we hope for a girl!"

Samson took Myna's chastel gently and put both to the side. He kissed her gently. He helped her off with her clothes and pushed her gently back onto the blanket.

His tool stood hard in open air. He lay down and bit gently on one of her nipples. He mounted and moved between her. He kissed both breasts and then went to her lips. He gently entered her tunnel.

Myna immediately jerked with the wetness and feeling of the first penetration. It was totally different, at least in her eyes, without a sheath! Her entire being flowed with the memory. Samson stood mildly surprised with the intense difference of going bare! He began the intense thrust that came with overpowering need! Then he let go and freely released!

Myna could feel the liquid physically secreting and slowly dripping out! But she knew that something intensely powerful still remained! She kissed Samson full on the lips and turned both of them to a side. She lay there warm in her husband's arms!

The troops arrived back at Village Littleton five days later! Sasselia immediately called Samson and Myna into her home.

"The council has made their final decision! They approve the pledge and arrangements will, indeed, be made to reactivate it! The next step would be to contact the Dwarven Council and arrange a Territorial Conference between both nations to sanctify the signing of said pledge! This will take time, though!"

"How long?"

"Theoretically, four to five years!"

"As soon as I return home I will inform the Dwarven Council! I had hoped it could be arranged for sooner!"

"With all the villages to inform and the arrangements to be made... a location found accessible to both nations... preparation and storage of food stuffs... alleviation of any internal stigma... on both parts... this does seem like an impossible task... but at least we got the ball rolling!"

"Then we have all we need! It's time my wife and I returned home!"

"Of course!" Then Sassy held her arms out for a hug! Samson complied brushing his whiskery beard against Sassy's baby-skin cheek. Myna smiled and hugged Sassy too!

"See you soon, sis!" she offered.

Anna and Margo stood holding hands, shedding tears on both parts! Margo touched Anna's nose with his index finger. He let go suddenly, walking toward the troops who were already mounted. Anna cried twice as hard as he mounted beside his parents. But Margo cleared his own tears in the knowledge that they would indeed see each other again! And some time soon, they both privately hoped!

Exuberant Confirmation!!

Samson convened the Dwarven Council shortly after he got back from the Elven Terriories! He asked all the current councilors as well as the retirees to attend this one meeting.

"I humbly admit this document into consideration." Samson opened. "I will pass it around the table and I freely open this case for discussion!"

The councilors, sitting and standing, each got a glance reading of the Pledge of Allegiance. Samson wondered if there would be mixed feelings about his grand risk! Peter spoke first.

"This is a good idea... I think! But the stigma of... let me guess... you're a sex-driven man! Your motive to this pledge is to promote interbreeding!"

"Bingo!" Samson admitted.

"The ethics of interrelations are kind of hard to get over!" Simon Sr. suggested. "For ages we Dwarves have held a code of ethics which restricted breeding within our own race!"

"Think of the prosperity of inter-territorial economics! It could blow our treasury off the top! This war has depleted our treasury! We need outside income, and we need it fast!" Peter argued. Samantha sat silently as she listened to her brother's argument. But Samson noticed that she held a sly look on her face.

"Sam, you have anything to say?" he asked. Samantha was reluctant to speak up... her own private thoughts were stigmatized.

"Making love with an Elf..." she finally relented. "Now, that sounds tremendously kinky!"

"But it's always possible!"

"You marry an Elf! I know it is, but that sounds like just the kind of wild act I would enjoy!" She smiled her best smile at Peter.

"Whatever turns you on!" Peter replied.

"Within limits!" Samson put in.

"What's your exact mission concerning said document?" Simon Sr. asked.

"My queen and I want to reactivate a pledge of allegiance holding our own signatures. Sasselia's and mine! The Elves have already agreed. All we need is a unanimous decision confirming the follow through of said pledge!" Suddenly he saw all hands rise. He had sincerely expected much more opposition! His council was truly made up of modern, realistically thinking youths.

"Then the follow through will continue!"

"How long will it take?" Peter asked.

"Four years at least! We have much to prepare! But, beginning now, I'm encouraging mediation between our nations! There will be many transactions between our peoples from now until this pledge is indeed sanctified!"

"And both nations will prosper!" Peter concluded.

Sassy's Letter

Samson shuffled through the five envelopes, his morning mail, which Sanders, his assigned scribe, had brought to him. He checked the seal of each of them but only one of them caught his eye. The Fireball!

Samson turned it over and saw that the letter had come from Sasselia Littleton! He immediately broke the seal and opened the letter.

Dear Samson: I regret to inform you that Alvin has been re-enlisted in the Elven Guard! We were planning a second child but he suddenly got notice! He left immediately! I do imagine I won't see him for quite some time!

The Elven Guard has been having problems with piracy! A group of chaotic thieves have taken to sea! The word is that the Colina thieves have also moved to Anasthasia and have joined in the operation! This war is being called The Pirate War by Anasthasius! Lord knows how long he will be gone!

Alvin promised a second child when he gets back... maybe he will get leave... but he may never come back! Piracy is serious shit!! I only hope....

Sassy had left the letter unfinished. Samson knew that it had been done on purpose. He conjectured on what step to take next. He decided to show the letter to his wife.

Myna sat sewing in her office. Samson approached and put the parchment in front of her face. She took it and read it twice. Then she frowned in concern. She turned in her chair and raised her eyes to his.

"This is a problem! Shame on Alvin for leaving her barren!"

"I wish I could help!" Samson confessed. Myna shrugged.

"This is a Principality problem! It has little to do with Dwarven politics!"

"Except that my fellow Avenger is dying of loneliness!"

"There's nothing you can do! Not right now, anyway!"

"I feel so useless! I wish I could at least offer her some comfort!"

"You will! In time!"

"What's that mean?" Myna kept mute and just shrugged her shoulders in silence.

Serious Discussions

The war had lasted six months now! And there was still no end in sight! The more pirates that were slain the more that enlisted! In the past six months Alvin had not returned home once to fulfill the conception!

Samson sat at home, looking at his latest statistics on Dwarven wealth! Sanders approached.

"There is a gent outside who wants to see you!" he opened. "He says he's from the Elven Guard!" Samson tore outside in a rush. He passed by Sanders in the doorway and approached a leather armored Elven character on horse back. But his hopes were diminished. It was not Alvin!

The guard stepped down and approached. He knelt at Samson's feet. He looked scraggly with a growing beard and had not slept his entire voyage, Samson could see.

"I greet you, your highness! My name is Saman Guard. I bring very dire news!"

"Alvin Guard is dead!" Samson hypothesized. The man nodded.

"He died honorably!" Samson turned red with rage. He had known this would be Alvin's destiny! Samson had experienced a sick feeling of edge for the past six months. Ever since the war broke out!

"Does Sassy know?"

"Leo himself asks you to break the news for him! We have the body in safe keeping in Anasthasia Morgue! He asks that you personally bring Alvin home! He has been dead two weeks now, but we did get our revenge! His assailant lies in the morgue next to him!" Samson smiled his good humor despite the situation. He returned to the house, welcoming his guest, and offering bath and bed. Then he went directly to Myna.

He moved to a drawer and retrieved his platinum dagger which he had left there for safe keeping. Undoing his belt he connected the sheath and made sure the blade was secure. Myna remained deathly silent.

"Alvin Littleton is dead!" Samson exclaimed. "In the morning I leave with Saman for Anasthasia. I will at least take the honor of guiding his corpse safely home. But... I have to face it! This won't exactly bruise Sassy's ego, but she's seen enough corpses in her day!" Samson stood with his head down in contemplation.

"I guess it's up to you!" Myna suggested.

"What's up to me?"

"Sassy's pregnancy, of course!" Samson turned physically white in confusion. Myna took Samson's hand and placed it on her fertile womb.

"How many times have you made love wishing I were Sassy?" she asked. Samson stood dumbfounded. "I'm willing to accept Sassy as your second wife! After all, she's my sister!" Myna held Samson's hand to her bloating womb. He could feel the child's heartbeat within.

Myna only had three months left to go and Samson was contemplating leaving! He vowed to be back for the birthing! "I want to show you something! Myna walked up to her sewing area, opened a drawer, and took out two broken envelopes. She handed both to Samson. "Sassy has been writing me! I want you to keep these!" Samson read both parchments.

Dear spiritual sis:

I'm writing you because I realize, now, what I have been missing! Alvin has not been an exuberant husband! In fact, I would have had a better life with Samson! I could of learned to love him, and right now I find that I crave his attention!

I have been so blind all these years! Samson has treasured me, through thick and thin, and now, with Alvin's absence, I realize how much I crave loving attention! I think that you are the luckiest woman alive! This is my confession to you! Princess Sasselia Littleton August 24, 1376

Dear Myna: I have been alone in the Territories for four months now! Anna has become an acolyte and is currently living in the Elven Temple. How much I crave love! But, unfortunately, the love I crave is sexual!

I do want a second child, desperately, but my lover is unavailable. I wish I had chosen differently! God, please give me a second chance! Princess Sasselia Littleton October 31, 1376

Samson held tears in his eyes. He grabbed his pregnant wife hard and stormed his anguish.

"I do love you!" he swore. "But unfortunately, I'm caught in a triangle!"

"Go with it!" Myna suggested. "Don't fight your feelings!" Then Myna moved away to arm's length and clinged hands with him. "Give Sassy her second chance!"

Samson broke in a smile and then stormed his laughter as he clung to his understanding first wife!

The Direct Result Of Piracy Finally Paradise!!

Six days had passed. Samson and Saman stood at the edge of the raging river and stared across. On the other side stood Anasthasia Waterfront! Samson walked closer to the edge and strode aboard the small ferry which would take them across. The crossing went without incident.

It was storming rain when Samson set first foot in the City Morgue. Classic atmosphere for a burial!

Samson was referred to the city coroner. He brought Samson to a room which was called, simply, "The Vault."

"We need a positive identity for the papers! Basic procedure!" The coroner crossed The Vault and stopped on the far side of the room. The temperature was cool, Samson noticed. Then the coroner opened a cabinet labeled simply,

Alvin Guard

The drawer opened and cold steam rose. Then drawer contained a wooden coffin. The lid stood to the side. The drawer lay extended. A body lay inside, covered by a leather wrapping! The leather was moved aside and Samson saw the corpse. It was, indeed, the body of Alvin Littleton!

"Punctured lung, completely lacerated small intestine! Internal bleeding did the rest!"

Samson saw Saman suddenly standing next to him. Saman crossed his heart.

"Birth name?" the coroner asked.

"Alvin of Village Littleton! Headman of said village!" Samson verified.

"Thank you. He will be listed as Alvin Littleton!" Saman held tears in his eyes.

"Alvin Guard. That was his little pun! 'I'm Alvin Guard of the Elven Guard' he used to joke."

"You can't take him by sea!" the coroner suggested.

"We go by land!" Samson insisted.

"I've arranged for a horse and buggy for the trip." Saman spoke up. Samson nodded. The coffin was put on a dolly and they began Alvin's grand journey home.

Samson walked out of the morgue with a hand in a handle supporting the coffin with Saman behind. He suddenly noticed that the entire Elven Guard stood in attendance on either side of the entrance. One attendant approached.

"I'm Captain Serinus Guard of the Elven Saint! My condolences to the Princess!" Then he took the handle from Samson and helped Saman load it up.

Samson had estimated that by horse and buggy it would take 12 days to cross the continent! But he planned to travel non-stop. The body would slowly decompose with exposure! Samson made sure it was not too bad.

Samson arrived in Village Littleton driving horse and buggy, and every member of the village saw the coffin. Samson stopped in front of the Littleton residence.

People surrounded the coffin in question. They all knew death. It was all too usual. But who?

Samson knocked on the door. Sassy answered slowly. Then she saw the visitor. But her eyes suddenly lay on the buggy... and the coffin lying in the back! She ran past Samson and forcefully removed the lid! She saw the corpse.

She had known all along! She knew that Alvin would not be coming home alive! She turned away from the sight and knelt in anguish against the wheel of the buggy.

Samson stood quiet. He knelt beside her and offered an arm behind her back. Sassy had fought next to Samson, had seen how a corpse will decompose, and knew that this was second-rate.

She turned to Samson and offered her nose. They played a series of affectionate looks but she suddenly pulled away. Someone enquired what must be done. Sassy went to work preparing arrangements of the funeral!

The coffin holding the remains of Alvin Littleton had been put in temporary storage. The funeral would be held the next day.

Sassy sat in her living room next to Samson. She thoroughly enjoyed Samson's company but wondered exactly where this would go.

"How long you staying?" she asked. Samson crossed a finger across his nose and hypothesized.

"A couple of months!" he suddenly exclaimed. Sassy stood dumbfounded in shock.

"That long?"

"You need me! In more ways than one!" he offered.

"Your wife needs you too!" she reminded. Samson sided that and reached in his leather jacket. He handed her two opened envelopes. Sassy looked! Then she looked again! She was suddenly on her feet, red with shame!

"I've betrayed you!" she answered.

"You have not!" Samson answered calmly.

"I bled my heart in these!" she suggested.

"And I'm grateful!" Samson replied.

"What do you want from me?" Sassy asked in fear of her own feelings.

"I want your love!" Samson offered. He rose and joined her, playing with her fingers.

"I want you! I want your children! And you want mine, I believe!"

"It's impossible!" Sassy whispered.

"Is it really?"

"You're already taken!"

"You mean a king of the Dwarven Territories can't have two wives?" Sassy laughed her confusion.

"You actually want me to be your wife!"

"Second wife!"

"Why?"

" Because, my love, I know how lonely you have been for the past six months!"

"I don't know!" Sassy wondered.

"Let's at least try!"

Sassy could no longer resist! She physically lowered her head and grasped Samson in a kiss!

Samson responded. Sassy found herself intensely interested. She immediately dropped to the middle of the living room floor. Samson was next to her. He removed most of his clothes. Sassy let him remover her own.

Sassy's breasts were ample, more than ample, and Samson realized that he was wild with excitement.

He wanted it to last! But her beauty was beyond compare! Sassy bent down to him and kissed him full! Samson noticed that, except for leg length, they were adequately positioned.

Samson moved a hand to her legs and fondled her. She watered! He lay on her side and kissed her back! Then he could no longer resist! He entered from the front.

Sassy gasped her pleasure! The entry was a flush of revelation for her. He moved in rhythm with her and saw she was responding well. She gushed water again! Samson kissed her gently and bit one of her nipples. Then he suddenly pumped deep and released. Sassy yelped her surprise in uncertainty! But it melted away with Samson's comforting presence!

Confession And Burial

Sassy awoke feeling totally rested! She turned her head to see the man snoring next to her. Then she remembered. She smiled her glee.

She dressed and strode into the living room. Samantha Littleton stood at the door. She walked in and knocked on the wall. Sassy freely greeted her with a hug.

Samantha looked into Sassy's eyes. She saw joy for the first time since Sassy's matrimony and Anna's birth! She hugged her younger half sister and they sat.

"What is it that has you so pleased?" she asked. Sassy took Sam into the bedroom and pointed out the man sleeping there.

"King Samson himself?" Sassy nodded. They left Samson to sleep. Sam looked into Sassy's eyes again... she saw it... the intense joy of sexual gratification! This took Sam by surprise.

"You mean... you and Samson..." Sassy laughed out loud. "How did it feel?"

"The same! Only there was more water! It's amazing how two totally different races can feel the same!"

"So Samson's sexual acts were..."

"Absolutely the same! It's as if I never lost Alvin! And, this time, I know he loves me!"

"So how long..."

"Just last night! You know how long I've wanted to have a second child!"

"It's possible?"

"Absolutely! I have faith that this child will develop adequately!"

"But it will certainly be different!"

"Only in height! Dwarves and Elves are distant cousins, after all!"

"According to the Pledge of Allegiance!"

"We must hold faith in said pledge."

"Maybe... the headmen said four years to arrange a signing? This pregnancy may prove to push the signing up!" Sassy freely nodded.

Anna Littleton, Sassy's only child, came running into the house yelling at the top of her lungs.

"Mommy!" Then she ran directly into Sassy's arms.

Ludwig Wildlife himself, chief headman of all the Elven Territories, strode in behind Anna and moved directly toward Sassy's bedroom. Sassy shook with physical fear.

Ludwig sat himself next to Samson's bed and shook him awake! Samson wakened immediately. Then he stared!

"Welcome, my son!" Ludwig greeted. Samson stood at a total loss for words. Sassy appeared in the doorway.

"Sassy, it's about time you acknowledged Samson's love!"

"What?" Sassy moved to join them.

"I knew that as soon as Alvin died you'd run to Samson! Samson is one who has always loved you!"

"How...?" Samson asked.

"Exceptor told me! During the Wildlife Campaign!"

"Yet you let nature take its course!"

"Royalty begets royalty! After I learned of your heritage I realized that..." he pointed at Samson. "Samson Trellis was the man for her! And I also realized, a long time ago, that it would indeed lead to fertility and a child!" Samson stood spellbound. "In some ways the connection between Exceptor and myself has never been totally severed." Samson smiled.

"My daughter has been having a very rough time! I thank you for nourishing her love!"

"I love her as a wife!" Samson freely admitted.

"I know you do! And for that I pronounce to push up the signing! After all, this child will be considered illegitimate! The signing will, indeed, sanctify more than just a pledge! It will sanctify this child's creation! Each Dwelf born is a mascot of this pledge!"

"It's time we held the funeral!" Sassy informed. Samson nodded and Ludwig left to give him privacy.

The funeral would be held in the Elven Temple. A stage had been put up and Alvin's coffin, and its celebrated contents, lay against the far wall. The lid remained sealed.

The village assembled. Seats were taken but the acolytes all made a row of lines to the side... including Anna Littleton!

Samantha stood up to the stage and addressed the audience.

"We gather here today to celebrate the life of our beloved brother, Alvin Littleton. And to coronate his remains into the Underworld! We begin with his eulogy, conducted by his widow, Sasselia Wildlife!"

Conversation spread at the name 'Wildlife'. Apparently, Sassy had renounced her married name! Sassy took the stage.

"I'm afraid that this eulogy will not be a nice one! Alvin Littleton was a soldier! He was totally dedicated to the sailing police named the Elven Guard! So dedicated that he forsook his position as Headman of this village, he forsook his so-called marriage, and he betrayed his own daughter!"

"Alvin had little, if anything, to do with Anna's raising! He avoided us both as often as possible! His betrayal and lack of attention to priority is what killed him... not any enemy he may have faced! As Princess of the Elven Territories I reclaim my maiden name and disown the name 'Littleton'.

"Further, my daughter will me renamed Anna Wildlife! My devotion to Alvin was always a lost cause! My dedication to his well-being was shrugged, cursed, and ignored to eternity. May the Underworld treat him mercifully!"

"My next proclamation as Princess is the signing of the Pledge of Allegiance! Which will become permanent legislation!"

"We, the mothers of the new race named Dwelf, will stand together and enjoy our union with our Dwarven lovers!"

"Furthermore, I, as Samson Trellis' lover, hope to conceive a child of my own! I will become second wife and royal liege to this pledge! I humbly relent Alvin's remains to the Underworld with love and understanding!"

Sassy walked off the stage in satisfaction. Samantha took the stage back and began the coronation rituals.

She waved her hands, creating a fireball, and turned throwing it directly onto the top of the coffin. By some ominous coincidence the lid suddenly fell in. Then each acolyte threw a just-learned fireball on the outside. The remains would be instantly cremated.

After the fireballs were all dispersed, all that remained was a pile of small wood chips surrounding the burning remains of the corpse.

The fire would burn until the morning after and the building was made of pure stone, so there would be no fire hazard!

The room cleared. People stood gossiping about the pronouncements Sassy had spoke, but she ignored the gossips as Samson took her in hand to walk her home!

Requesting Consent

Samson and Sassy walked into the house. Samson immediately sat down taking Sassy with him.

"What was Alvin really like?" he asked.

"You heard the eulogy!"

"He must have had some positives." Sassy sat back and thought about it.

"Well, we liked each other as kids! We were promised at birth! Our parents nurtured our relationship hoping we would bond! And we did, in some ways! His favorite game was Knight in Shining Armor! I was the damsel in distress, of course!"

"I had hoped that it would work... I really had! But I knew Alvin's character from the start. He always wanted to prove he was the macho hero! That's why he joined the Elven Guard!"

"Actually, he had very little self-esteem! That's why he always felt he had to prove himself! Anna's conception was his way of proving he was a man! But after her birth as she grew he drifted away! Began dreaming of and missing his Elven friends. When he became headman he did not want the position! But he didn't refuse, either!"

"Unfortunately, his marriage to me promoted him to headman automatically! He shrugged his shoulders and went with it... but in some ways he always ignored the responsibility! He was the same with Anna too! After the first few times of rejection Anna just gave up and gave all her love to me! And I do thank her for it!"

"What happens if you do conceive? How will Anna react?"

"She's always wanted a sibling!"

"I mean," Samson responded, "how will she react to her half-brother's father also being her father-in-law?" Sassy thought that one over.

"I hadn't thought on that one!" Sassy confessed.

"We have to talk to her!"

"She might need some counseling anyway!" Sassy thought. Samson rose and gestured to her to rise with him. He took her by the hand and led her across the room.

"Let's go talk to her now!" Sassy freely agreed.

They walked into the residential area of the Elven Temple! Then they knocked on Anna's door. It opened immediately. Anna stood staring at her visitors.

"Can we come in?"

"Sure!" Anna moved aside.

Samson and Sassy walked in hand-in-hand. Anna sat down and waited politely for them to speak. Samson grabbed a chair and sat across from her.

"Anna, do you understand how complicated life can be?"

"Very! Now that my father's gone mom is all I have left."

"That's not true!" Samson assured. "You have me, you have Aunt Myna and, of course, you have Margo!"

"I heard that you two want to marry!" Anna spilled the beans.

"Yes we do!" Samson confessed.

"How's that affect me?"

"I'll be a step father as well as a father-in-law."

"That's confusing!"

"You realize that family units are often complicated! But I want your permission to marry your mother!"

"What's the dif?"

"I need to know that you understand!" Samson assured.

"You love her?"

"She loves me!"

"That's not what I asked!" Samson smiled his surprise.

"I've loved her since the Wildlife Campaign... maybe even before!"

"Then why did you marry Myna?"

"Your mother was already promised! And the marriage between me and Myna was sort of arranged!"

"So you loved mommy first!"

"Yes, she would have been my first choice!"

"She was already promised... and there was no end to her marriage in sight! I think I understand!"

"Unfortunately, I'm afraid I've gotten myself into a love triangle!" Samson relented.

"Happens all too often!" Anna assured. "Mom has a triangle of her own! When she has been lonely she has said that she had married you! But that's recently!"

"I know how lonely your mom has been since Alvin left! And, to tell you the truth, it hurt me too!"

"My father didn't love anyone! He was too self-centered."

"You don't hate him, do you?"

"No, but I don't have any love for him, either!"

"He's traumatized you."

"In some ways! But there are always others to love!" Samson nodded his understanding.

"That Anna is one smart cookie!" Samson motioned as they left the temple.

"She's got her mother's genius!"

"I'm glad we talked to her!"

"This brings back memories! Alvin walked out on her the last time he rejected her! She literally cried tears of grief! But I comforted her and told her it would be okay! There were always others who love you! I'm afraid that line has stuck with her since!"

"It's something that sticks in the memory!"

They remained silent holding hands all the rest of the way to the house. Samson closed and barred the door!

"Let's do that living room thing again!" Sassy took her shirt off... she was not wearing anything underneath! Samson gawked his surprise. He instantly rose.

Sassy moved to him and brushed her nipples against his chest. He removed his leather shirt and moved into the living room. Then he took a blanket from the bed, spread it out and laid on it. Sassy joined him. Samson reached and pinched a nipple. Then he began to suck.

Sassy put a hand behind Samson's neck and pushed him closer. He moved between her breasts and began to nuzzle, slowly pushing her into a laying position. Then he removed her leather trousers and noticed again... no coverings! He removed his pants as she caressed his hairy chest!

Samson jumped on her and pushed her down! Then he surprised her by immediately entering.

Sassy gasped with pleasure as Samson pushed and pulled. Then she moved onto her stomach... he took her from behind!

Due to the night before Samson was taking his time. They played foreplay and enjoyed each other's bodies for an hour! Then he got serious!

Taking her from the back he closed her legs, pressurizing his cleft. Then he raised his chest to add shoulder pressure. He smoothed her clit with a finger. She watered and that very act brought him to climax! Samson lay exhausted letting his fluid flow into her freely!

Sentimental Moments

The next day Samson called for Anna to visit. She came in the early afternoon. Anna freely sat down next to Samson. Sassy sat on her opposite side.

"Anna, do you realize what the union between me and your mother means?" Samson asked.

"I understand."

"We want children together!"

"Do you really? I'd have a sibling?"

"Would you like that?"

"I'd love it! But make sure it's a boy! A younger sister would constantly contradict me!"

"We'll try! But, Anna, you have to know... we're already trying!" This took Anna by surprise. She moved away to give herself space.

"So... she may..."

"Yes, it's possible, she may have already conceived!" Anna turned white with concern.

"In that case... this child will be..."

"Illegitimate!" Sassy admitted.

"Ooo...kay! So, how about..."

"It may already be too late!" Anna shrugged her shoulders in relent.

"What am I worried about? I'm just being prejudiced!"

"Anna... we promise, when you grow up, we won't censor your love life!" This seemed to calm her down.

"If you have a girl I'll never speak to you again!"

"I can't make any guarantees... nobody can!" Anna whined her concern. "I'm not about to abort this baby!" Sassy promised.

"I understand!"

"Can you accept it?" Samson asked.

"I can! The nature of this pregnancy is not my concern!"

"It's ours! But there will still be a possibility of stigma!"

"Why?"

"The mixture race of Elf and Dwarf will be called 'Dwelf'. The stigma of such race is already being worked on! Now, there are on record, already five Dwelf conceptions in these territories."

"Including my mother's?"

"Yes!"

"Why?"

"Anna," Sassy explained, "what's happening here is a revolution caused by the revelation of the Elven-Dwarven Pledge of Allegiance!"

"A pledge which was held top secret, I may add!" Anna revealed.

"Samson felt..."

"It was an easy way into your pants!" Anna frightfully suggested.

"Anna, I've always loved your mother!" Samson freely admitted. "Race, creed or make don't and shouldn't limit growing love! When you grow up you will marry whoever you fall in love with! Despite all these!"

"I thought I'd marry Margo!"

"If that's still your wish! We shall wait and see!"

"So this promise isn't official!"

"It never was!" Sassy and Samson both agreed.

"My marriage to Alvin was arranged... look what happened to it!"

"So we wait and see!"

"On both parts!" Sassy admitted. Anna sat back down. Then she asked another question.

"Why now?"

"Why what now?"

"Why reveal this pledge now?"

"Why not? This pledge has been classified and top secret for fifty years! It was official legislation before that!"

"It was?" Samson pulled out his copy. Anna read it.

"1326!"

"Fifty years ago!"

"Why'd they classify it?" Anna asked.

"No one knows!" Sassy revealed. "No one living at least!"

"Is there a way to find out?"

"I've assigned two scribes to locate and translate any and all documents written by either Horticus Trellis or Isidius Wildlife!"

"So it's possible to find out!"

"We hope!"

"I'm sorry! I was totally uneducated."

"Now you are!"

"How long before you know?"

"About which?" Anna stood up.

"The baby, of course!" she exclaimed.

"I'm waiting another week! If I miss my period until... Friday, then it's confirmed!"

"I'm certainly hopeful it won't come! My brother may already be on his way!" She skipped out the living room and out the front door. Sassy rested her head on Samson's shoulder.

"That was the toughest time I've had yet!" she admitted. "Much harder than Alvin's passing!"

The assigned Friday came. But Sassy's period hadn't! She was at the sink washing dishes. Samson approached and wrapped his arms around her.

"It's official! You were as active last night as before!"

"I saw Tralina yesterday!" Sassy informed. "She says the possibility is confirmed. I'm pregnant!"

"I thought so!" Samson moved away and began to sharpen his dagger.

"There's something else!" Sassy revealed. "I saw the scribes today!"

"And?"

"They gave me a roll of parchments which Isidius wrote!" She gave the parchments to Samson. "I think Anna should be here!" Samson nodded.

Anna, Sassy and Samson had gathered. Sassy opened the roll. Then she turned to the first page.

"June 18, 1326. Horticus Trellis is dead! He died June 9th. He leaves behind Andrew Trellis, who will sit as new king, the illegitimate son Henry Tralesta whose mother is Asandra Trellis, and his Dwarven wife, Isrialda Trellis! The following documents describe our shared family trees!

Sassy turned to Anna.

"According to these records Henry Tralesta is the direct ancestor of Anna Peesar Tralesta, Tralina's grandmother! I will have to notify Tralina about this!"

"So Tralina is descended from the Trellis line also!"

"From a distance, yes! She's also my first cousin! There's more! After the family trees there's one more parchment."

"August 31, 1326. Asandra gave birth to Horticus' second child on June 11, 1326! She had been expecting when he died! She grieves his passing but relents to it with his extra gift!"

"September 10, 1326. Henry Tralesta has murdered his own mother! The stigma against Dwelves has been intense! Henry has sat on the council of headmen for two years! He requests that the Pledge of Allegiance be disbanded and held top secret."

"December 31. New Years Eve. The Pledge of Allegiance has been classified as top secret. New legislation is intact regarding spread of stigma against inter racial breeding! Some suggest stigmatizing reproductive factors!"

"I have been fighting on the council for the past months to reactively my pledge! But now, I see, I am nearing my demise! After I am gone the Pledge will, indeed, be old news! And in the past! I certainly hope..." Sassy was in tears. "that this pledge will present itself to someone who will accept our beliefs in equality of reproductive breeds and this very important document! This will be my last entry!" Isidius Wildlife July 20, 1327

Sassy cleared her tears. "My own beloved grandfather! I never knew him! He was Ludwig's father! We must follow through on this! It's my own grandfather's last request!"

"We will! This document will sanctify the pledge all the more!" Sassy nodded her tears away.

"Think, this child will be born an honor to Isidius!" Anna suggested.

"Then let's name it after him!"

"If it's a boy!" Sassy agreed.

"And a girl?"

"Asandra, after Hortius' lover!"

Stewart Wildlife

Two more weeks passed without finding any more information. Samson knew that he would have to return home soon! He packed his bags, and as he was working, a scribe suddenly approached.

"I greet you, your highness! I humbly request a presence with both you and the Princess before you depart!"

"On what subject, may I ask?"

"I carry documentation on Dwelven Culture!"

"Dwelven Culture?"

"You may be interested to know that Dwelven society still exists!"

"Indeed! Come! We will attend you immediately!" Samson led the way into the house. Sassy was sitting with Anna in her living room. Samson called Sassy to attend. Anna got up and headed toward the door.

"I greet you, my lady!" the scribe opened. "I humbly request to aid you as personal scribe."

"I already have a personal scribe! Give me good reason to replace him!"

The scribe was dressed, like all learned scribes, in a cloak with a hood. It was custom that scribes were never asked to identify themselves! They were considered learned and mystical teachers! This man suddenly lowered the hood from around his face, which was considered an honor seldom given.

"My name is Stewart Wildlife... I was born a Dwelf!" Sassy gasped in surprise.

"You claim to come from Village Wildlife. Which line are you from?"

"I am the Dwelven son of Cassandra of Claremill Settlement and your grandfather, Isidius Wildlife!"

"You must be joking!"

"I assure you, I'm not! I was born your Dwelven uncle!" Samson smiled in sudden comprehension.

"Sassy, if Horticus had an elven mate, then wouldn't Isidius choose a Dwarven one?" Sassy strode around Stewart in mild curiosity.

Stewart stood at 5' 7" which was three inches shorter than Sassy and three inches taller than Samson. He looked just like any other Elven or Dwarven character. His height was the only difference! Sassy was finally satisfied.

"Your offspring will be one of us!" Sassy walked around him again.

"How old are you?"

"I was born August 1, 1316."

"That makes you sixty."

"That is correct!"

"Why do you want to work for me? Because of our bloodline?"

"I have waited 34 years to make your acquaintance."

"That's my entire lifetime!"

"I heard of your birth shortly afterward."

"Why haven't we met before?"

"Up until ten years ago I was in hiding!"

"From whom?"

"Henry Tralesta, of course!"

"You knew him?" both asked.

"Isidius did! He feared for my life! He sent me away when I was ten!"

"Where'd you go?"

"To Claremill Settlement where my mother grew up!"

"Of course!"

"Claremill was one of the first settlements ransacked during the war! Everyone escaped safely!"

"That's good!"

"I have also taken part in documenting the Dwarven War!"

"Indeed!" Samson seemed impressed.

"So you returned home at the age of fifty! Why?"

"Because I had heard, finally, of Henry Tralesta's demise!"

"He killed his own mother!"

"It was an act of rage!" Stewart revealed. "He had not been told about his lineage! That was Asandra's doom! He felt that Dwelves were a creation of incest! He looked at the pledge and its results in just that way!"

"So how many of you are there?"

"Twenty survivors! We all became scribes!"

"Why?"

"To hide our identities! After all, a scribe is never asked to reveal himself!" Sassy nodded her comprehension. "I have something for you!" He reached in his leather bag and took out a roll of parchments. "This is my own work covering the occurrences within these territories these past fifty years! It covers all twenty survivors and their lineages! Also, we included Dwelven traditions!" He handed the roll over which weighed in hand. "Keep it safe!"

"We will!"

"I will go now! Your majesty, I hope you enjoy your trip home!" Stewart strode out. Sassy put the roll away then turned back to Samson.

"I have to talk to you!"

"Feel free!"

"I talked to Anna!"

"I noticed."

"I asked her if she wanted to change her name! She declined saying her father was her father! She still wants to give him his rightful honor!"

"That's understandable!"

"I will consent, of course! It's something she can give him with what love she does have for him!"

Samson sat astride Chestnut which he was borrowing for the trip home! They had spent the last two hours just smooching and enjoying each other's company. Samson gestured a salute and started his journey home. But, for some reason, he looked forward to Myna's attention! He knew that she would go into labor very soon! He had to get back and so he would!

The New Year Passes

Samson arrived at Trellis Settlement. He unbridled Chestnut and put her in her stall. When he was finished he turned toward the house. Myna was standing in the stable doorway.

"I missed you!" she greeted. Samson smiled his joy and embraced her kissing her intensely. "Unfortunately, I'm in no shape for it!" Myna reminded. Samson understood her regret.

"That's okay!"

"I've talked to Margo about Sassy! He relented to the joining with some curiosity!"

"I still want to talk to him!"

"That's fine!"

After they entered the house Samson saw Margo run into his arms. Margo was now aged at 7.

"Hello, you big stag!" Samson laughed. Then he sat seating his son next to him.

"You know that Sassy's expecting?"

"Is it yours?"

"It certainly is!"

"Then I will have a half-sibling!"

"As well as a full blooded!" Samson reminded.

"Does that right come from your title?"

"It's not just my lineage! The pledge states that each man may have two wives... one elven, one dwarves! But only those two!"

"Then it has a limit!"

"They're both more than I can handle!"

"Anna still wants me?"

"She certainly does! But that's still to come!"

"Mother's entering her eighth month!"

"I know! Don't be surprised if Anna arrives early!"

"You named her Palapany! What happens if it turns out to be a boy?"

"It won't!"

"You seem so sure!"

"The centaur is a spiritual being! It can see the future! I'm sure and your mother is too!"

"Then it's a girl!"

Alanah entered and reassured that the pregnancy was proceeding as expected. Then she left. Samson ruffled his son's black hair.

"You'll have Dwelven children yourself!" he assured.

"I know!"

"I must look into other things now! Dwarven politics still continues!" Samson rose and moved to his office to do some much needed work.

Myna went into labor on November 25, 1376. She did not give birth until the next day. Anna Palapany Trellis was born November 26, 1376 at 10:04 a.m.

The new year arrived and went. Politics and life in the Territories continued as usual.

Six months passed. On April 10th Samson was working in his shop! He suddenly heard the heavy trot of multiple hooves! He looked up to see a troop of horses approaching.

Sasselia Wildlife sat in the lead. She looked very beautiful to him! Samson took pregnancy to be a beautiful and mysterious process!

He ran to see her and helped her dismount. She was heavy with child... maybe too heavy! Then he saw Tralina Tralesta riding in back!

Samson greeted his guests and stabled the horses. Then he entered the house and saw Tralina in consultation with both Myna and Alanah! Sassy came to him and they both took seats.

"We have to talk!"

"Sure!"

"I've been passing blood! This pregnancy seems different in some strange way!"

"Are you okay?"

"Passing blood is not necessarily dangerous. But Tralina wants others to check me out."

"Then lets!"

Sassy lay on a cot as Alanah and Myna examined her. Alanah put a stethoscope to Sassy's bloated tummy. She turned and handed it to Myna. She listened too! The device fell from Myna's grasp! A consultation was called in the next room!

"You heard them?" Alanah asked.

"I did!"

"Two?" Tralina wondered. Myna laughed.

"Sassy is carrying twins!"

"You sure?"

"Positive! Two heart beats mean two babies!" Alanah assured.

"My mother knows how to handle a situation like this!" Myna assured. "Her own mother specialized in multiple births!"

"What do we do?"

"Wait! It's possible, if Sassy takes care of herself, despite the bleeding, she can still birth safely! I recommend that she stay here for the rest of the duration!"

"Agreed! Soon she will be in no shape to ride!" Myna insisted.

"Understood!"

"Twins!"

"Let's not tell them! We'll just console Sassy and reassure her."

Six month old Anna Trellis was brought from her crib and placed in Sassy's arms. Sassy smiled her glory and noticed her likeness to both her progenitors!

"We nicknamed her Palapany!"

The midwives returned and stated that according to their findings everything seemed to be running smoothly despite the bleeding! Sassy just nodded, captured by the beauty she held in her arms!

Myna and Alanah exchanged knowing glances. Alanah wanted to break out in laughter but instead she just left the room!

Sassy was finally free to rise from her cot. She handed Palapany to Myna and rose. She went to Samson and gestured an okay. Samson smiled his relief. Then Sassy sat next to him.

"You make beautiful babies!" Sassy complimented. Then she turned to face him. "I have notice on the signing! It's set for July 20th."

"What year?" Samson wondered.

"This year! July coming up! My only concern is my labor! I hope the child will come early since I'm stuck here until then!"

"July 20th, 1377?"

"Yes!" Samson shook with surprise. So soon!

"Any details?"

"All of them have been arranged! Fortunately, my father has influence on the council!"

"I'll say!"

"Our matrimony will be held on the last day of the three day festival! It's being named The Pledge Festival! They plan to hold one each year! And, the location for each festival has been chosen! On the outskirts of the Elven Forest where there's grassland and farm settlements! This festival will obliterate exactly fifty years of secrecy concerning the pledge." Samson nodded.

Twin Births

The month of June arrived. Sassy was expecting any day now, but still carried. In past months she had grown to think of Trellis Settlement as a second home. She now felt very attached to Palapany... just like a second mom.

One day Sassy entered the living room. She noticed Samson lounging on the couch. She sat next to him and turned to him.

"You do realize that it will be you and Ludwig signing the pledge!"

"I did realize this as an afterthought." he informed. "After all, he is the god head of those territories."

On the day of the 11th of June Sassy sat lounging on the couch as she had just awakened. She realized it was time to search for some breakfast. The question was, would she keep it down?

She rose slowly, in consideration of her condition, and slowly marched to the kitchen. She went into the cupboard to choose a plate... She felt the first pangs of labor so suddenly that she let go of the plate... it shattered as it hit the floor!

Alanah and Myna immediately appeared. Both saw the shattered glass which surrounded Sassy's feet. She buckled over in pain and physically belched.

Alanah ran to her and Myna helped move her toward a bed. Myna prepared hot water as Alanah continued to prep the bed.

"Go get Samson!" Alanah yelled her advice... Myna ran... the front door slammed behind her!

Sassy had been in labor for the past twenty hours... she was still nowhere near delivering!

Isidius was born at 10 pm. After his delivery Sassy was taken aback at her continued spasms. She wondered if she held the strength to continue! Alanah assured her that she still held deep reserves.

Samson heard the healthy cry of a new born infant! He waited for Alanah to reveal herself... time passed... she had not appeared and Samson grew nervous! He walked to the door and knocked softly! No answer! He knocked harder! Still no answer! He felt like storming his way in but Alanah suddenly revealed herself!

"I heard the cry of delivery... what's holding you up?"

"Isidius has come, but Asandra still has not delivered!"

"What?"

"Sassy is mothering twins!" Then she silently retreated once more.

Samson turned from the door and stood spell-bound. He stood stunned with this news!

Asandra delivered 2 hours later. Samson was finally allowed in the room!

"Don't disturb her! The labor was hard on her... perhaps harder than last time! She needs her rest!"

Samson sat near Sassy's bed... she was out like a light in unconscious sleep! Samson grabbed Sassy's hand in his own and kissed her palm. Tears appeared in his eyes! He had decided that he would spend this night sleeping right next to her

Sassy spent all the next day in bed recovering. The day after she awoke feeling herself again. She turned and looked around... Myna stood at the crib. She suddenly turned... Myna held small tears in her eyes! Sassy smiled and cried tears of her own!

Samson entered the room followed by Alanah.

"I would like the birth names for the birth certificates!" Alanah informed.

"Asandra and Isidius Trellis!"

"Asandra and Isidius Wildlife!" Samson spoke at the same time. Everyone laughed.

"It seems we have to compromise on last names." Sassy noted. They finally decided on Isidius Trellis and Asandra Wildlife Trellis.

Sassy was up and around in due time! She had decided that she would leave for home on the 20th of the month. This way she could have a full month to prepare for the festival.

"I'm taking Isidius with me." Sassy informed. "It's a promise I made to Anna! You can keep Asandra here if you want... Myna has taken a liking to both of them!"

"I'll bring Asandra to the Pledge Festival... " Samson confirmed. "They can be reunited there."

June 20th came with a clear, sunny day! Sassy was mounted on Chestnut... she was taking her home. Samson smiled his goodbye as Sassy signed her departure. Tralina would care for Isidius as nanny. Sassy turned her horse around and took off toward the border. Samson turned to Myna and both walked back to the house.

Pledge Festival Is Born

The Dwarven and Elven Territories have announced a merger union based on the newly resurrected Pledge of Allegiance! The Annual Pledge Festival, which will indeed be a yearly event, will celebrate the union! It will cover a three day occurrence. Sasselia Wildlife and Samson Trellis, as royal lieges to said pledge, will be married in an Elven ceremony on the third day of the event!

Sassy has given birth to Samson's twin children, Isidius and Asandra. Isidius, of course, is named after the late Isidius Wildlife, while Asandra is named after Asandra Trellis, Horticus Trellis' Elven equal.

The third day of the Pledge Festival will involve the signing of the revised Pledge of Allegiance, which will become permanent legislation, by Samson Trellis and Ludwig Wildlife! As well, the matrimony will be held the same day! Daily Scribe and Weekly Writing

Pledge Festival

Day One!!

The opening day of the Annual Pledge Festival would be a day to arrive, get settled and enjoy yourself. Samson and Myna arrived on location at midday. Samson greeted Sassy with a hug and escorted Myna to help her set up camp!

The festival would consist of waterproof tents assembled all around a 20 acre lot! Samson and Myna were escorted to their quarters! Margo and Little Palapany settled on their own little corner. Aleesa would be settled in a separate camp along with Satira, Peter and Sam. Alsandra was taken by Alanah to see her half sister, Anna Littleton.

When things were arranged in the tent and everyone was settled Samson went to find Sassy, Anna and Asandra. He ended up at Sassy's headquarters! Anna had a smile on her face at the sight of Asandra.

"I guess I got a little bit of both sexes." she suggested.

"I suggest the infants be taken care of by nannies."

"We really should scout the camp!" Sassy agreed. She handed Asandra and Isidius into Tralina's care. Later on Myna would decide to visit and would stay to help care for the children.

The day was filled with joyous reunions, meeting old friends and making new ones!

Sassy noticed that, honor to her title, many people gawked at sight of her! When Samson joined her they were seen as the perfect couple... despite their height difference!

Myna also noticed the atmosphere... she had some concerns about it!

The day had passed and night had fallen! Myna and Samson both noticed the mix of Elven and Dwarven heights. This bunch basically looked like a normal crowd. The mix made for variety!

Samson spent most of the day with Sassy! But Myna did not let this atmosphere get to her. Tralina came into Myna's tent and declared that Samson would be spending the night with Sassy. Myna nodded consent, as of course, she had no choice! Tralina decided to stay with Myna overnight herself. She felt very sentimental toward Myna's case.

Day Two!!

Sassy was up early the next morning... she had much to attend to! She would be in meetings the whole day! Samson, surprisingly, chose to attend Myna for the day. Sassy would be occupied and he knew about and felt sentimental toward Myna's loneliness. This consoled her somewhat.

Sassy was attending to meetings, taking care of final arrangements, and generally very busy. She once returned to her tent and noticed twenty persons, scribes to be exact, waiting for her! She noticed Stewart Wildlife in their midst. He gestured a hello and spoke.

"Greetings, my liege. I introduce all twenty members of Dwelven society." Sassy stood impressed. Most of the Dwelves, she noted, were aged at fifty or older! Some had wives and children. Even grandchildren. Sassy cried open tears at the sight.

"You must all stand as honored guests!" Sassy insisted. "You will have a special place in the ceremonies! As you will always be honored in my heart!" Sassy assured with much sentimentality.

Myna and Samson had plenty of time to be affectionate... even to go so far as mating! But Myna still felt disturbed by the nature of the Pledge Festival's environment. Samson still chose to bed the night with Sassy.

Final Day!!

Early in the morning of the third day everyone enjoyed a free pancake breakfast! Things would be slow paced from then until noon!

Samson was once more forced to attend Sassy although he did take the time to have breakfast with his whole family!

Later that morning Myna observed Samson and Sassy discussing with a scribe. Myna hated herself for her secluded and isolated feelings! Despite her own policy of freedom concerning Samson's love life, she felt a sense of jealousy taking control! She knew she could no longer control these strong emotions!

As time passed the anguish and pain of feeling like an outsider was increasingly relevant! She wanted to hide her intense feelings, but Tralina, he constant companion, would not be fooled! She could read Myna's body language like the cover of a book. However, she chose not to interfere at this time!

At 7 in the evening the Pledge Ceremonies began. Samson and Sassy both took seats on a raised stage as all twenty Dwelven scribes stood to Sassy's side. Sassy convened the ceremonies.

"Greetings, fellow Elves, Dwarves and even Dwelves! I would like to begin these celebrations with a reading of the newly resurrected Pledge of Dwarven-Elven Allegiance. To officiate the reading of this document I introduce my own Dwelven uncle, Stewart Wildlife!" Stewart strode up to the front and faced the crowd.

"Before I start this reading, I just want to make a speech of my own! My name is Stewart Wildlife, and I was born the Dwelven son of Isidius Wildlife! I stand proud of my lineage and breeding! We Dwelves are not mutants... we are people, members of these Territories, just like you! And now, after fifty years it's about time the secrecy concerning this Pledge be lifted!" He received a piece of parchment from his friends and began to read.

"We, the Godheads of both the Elven and Dwarven Territories, hereby acknowledge our own lines of lineage and our joint relations! We are of one government in two bodies... we are one family! We are descended from similar ancestry and as descendents of our shared progenitors we hereby apply our personal seals and signatures authorizing permanent legislation concerning such pledge!"

Stewart moved back to his spot. Samson rose as Ludwig mounted the stage. Samson signed and applied his Eagle seal to the bottom of the pledge. Then Ludwig did the same with his Fireball seal! Sassy stood up and rolled up the scroll, sealing it with her own personal seal of office!

"I hereby acknowledge and seal the newly revised Pledge of Dwarven-Elven Allegiance!" She raised the sealed parchment for all to see! Applause broke out with cat calls and hoots. When the noise died down the party was moved a distance to a firepit standing ready but unlit!

Ludwig instructed Sassy and Samson to stand facing each other. A small jolt of grief suddenly stabbed Myna in the ribs! She could not believe, with all her hard effort, the pledge stood firm! But this matrimony gave her mixed feelings.

Sassy turned to face the pit and moved her hands! A fireball emerged which she cast directly into the pit full of dry wood! The intensity of the force's power caused five seconds of glorious light in this pitch black darkness! There was applause and awe at Sassy's ability.

Amanda Wildlife, head sorceress and mother to Sassy, as high priestess, would perform the ceremony! She wore ceremonial robes!

"We gather here today to sanctify the joining of Sasselia Littleton Wildlife and Samson Samuel Trellis under legislation of the Pledge of Allegiance. Samson Trellis, will you take this woman, Sasselia, as your honored liege and equal, respecting her and loving her as your Elven equal?"

Myna felt torn inside... Sassy was her husband's equal?

"I certainly will!" Samson noted Amanda's use of the word 'equal'. Myna would not be happy.

"High Princess Sasselia Wildlife, will you take this man, King Samson Samuel Trellis, as your fellow liege and Dwarven equal?"

"I certainly will!"

"Then circle the flames thrice!" Sassy and Samson circled together once!

"The power of sulfur for long lives!" Amanda spilled sulfur into the pit, which, naturally, made the flames brighter! They circled a second time!

"Power of coal to keep both mighty and strong!" The third circuit.

Amanda obtained a bowl containing liquid and poured.

"Power of blood to douse the flames!" The fire extinguished with this addition! "Sasselia Littleton Wildlife, Samson Samuel Trellis, you are now pronounced Dwarf and Elf. It is time for the gifts of sharing!"

Sassy reached back behind herself and revealed a sheathed long sword. She unsheated it!

"This sword is the very platinum sword which once contained the essence of one named 'Exceptor'. May you never have need of it." Samson smiled in glorious surprise! Sassy handed it back behind her! Samson reached at his belt and unsheathed a platinum dagger!

"I present this dagger, smithed by my own hand, the first piece successfully completed based on Platinum Smithing! I offer it in honor of Andrew Trellis!"

Myna shook with surprise... he had not even told her! Aleesa was heard crying real tears that her husband's work was not lost! Myna turned, Tralina with her, and they returned to camp! On the way Myna thought how... in the end... this ceremony had turned out in a positive way! Festival Reflections!

Samson stormed into Sassy's camp and pointed a finger at her!

"The next time your mother uses the word 'equal' in discussion I will personally ransack her quarters, obliterate her reputation and cast this humble pledge down the sewer!"

"What do you mean?" Sassy yelled back.

"I've been concerned about Myna's sanity. She has felt like an intruding third party! An outcast! I won't have that of my queen! My proposal to Myna was from my heart!" Samson informed. "She's the one who chose me! I asked her to join me... as my equal! And now Amanda has discredited that proposal!"

"I'm sorry! But this whole pledge is concerned with equality of breeds!" Sassy reminded softly.

"And I agree. But this matrimony is not about personal equality... my relationship with Myna is!"

"I understand."

"You are my liege and my equal under the legislated pledge! You have been, are now, and always will be my first choice for wife! But there must be balance between you and Myna has not been given the due respect!" Then he slowly walked out of camp.

Myna sat with Tralina next to her. She was deep in her own thoughts. She was strangely smiling her knowledge!

When Sassy had given birth to twins Myna had thought that it had emphasized Samson's love and had confirmed and sealed it... but now she knew that Sassy would spend much less time with him from now on! Myna herself was the ultimate champ! She was still Samson's first equal! They were both of the same race... the proposal had sealed their love... Sassy had not been proposed to! She also knew how Samson felt about their intimacy.

Samson suddenly appeared and immediately sat next to her! Tralina took that as a sign to depart.

"I'm sorry!" Myna suddenly broke out in laughter.

"I beat her!" Myna yelled. "I'm the first equal! She unknowingly challenged me, Queen of all the Territories, and lost! Amanda totally tore Sassy's chances apart!" Samson giggled his agreement.

"I'm glad you look at it that way."

"This festival was a scam; a political hoax! I hope next year's will be better!"

"I know how you have felt! You have felt the outsider! But the fact is, I never slept in Sassy's bed!"

"What?" Myna yelled.

"It was elementary that I had to join her but I did not sleep with her!"

"You are a man after my own heart!" Samson got up and searched in his pack. He revealed a roll of cloth which contained six fresh platinum daggers and he lined them up.

"I made these myself! I will dispurse them to each of my offspring! Each holds both of my family seals... the Eagle and the Falcon. I gave Sassy an exact replica."

"Very high quality!"

"By the way, I only had two sheaths left when we arrived! I used one already! I think it's time!" He showed her the sheath he held in his palm.

Myna and Samson both undressed and enjoyed each other's bodies afresh. Myna knew this would happen again... and again, and again, and again!

Epilogue

Samson would continue to rule the Dwarven Territories well into his nineties. Samson Trellis would die at the age of ninety-seven and Margo Trellis would become king after him!

Aleesa Trellis Anders would die at the age of ninety eight. The queen mother, Alanah Milner, would die at 73. Sonar Milner would join with Samantha Littleton and they would produce a second child for Sam's care.

Myna would outlive her husband and would nurture her son's leadership! The principality and both Territories would indeed continue to prosper!

Beyond Galala Valley The One Year Trip Into Unknown Territory

Surprising News

Margo Trellis was thirteen years old. He lived with King Samson and Myna, the Queen, as he was a prince. They lived in the Dwarven Territories, far from the Principality of the Crane, governed by Leo Anasthasius. The Dwarven Territories stood in the north east corner of the continent.

The Clan Wars were over and the Terrs had enjoyed seven years of solemn peace under Samson's care.

Samson stood shoeing a horse when the thought he heard the sound of shoe-pounding troops close by. He immediately stopped his work in the smithy and walked out into the clear, blue and sunny day.

The hooves continued louder. Who could be coming to visit?

Samson smiled when he saw the Tralesta family, Thonolan, Tralina and their eldest son, Thor. They were riding in style... in a coach.

Thor aged at 18 now, and he was growing into a strong and forceful thief!

"Well, look what the cat dragged in!" Samson joked.

"We have business to attend to!" Thonolan greeted.

"Come in the house! Make yourselves comfy!"

A steward helped the Tralestas down and prepared the horse for rest. Everyone took seats in the house.

"You said business?" Thonolan pressed a parchment into Samson's hand. Samson studied it. It held the Fireball Seal.

"This is incredible! A troop of youngsters is assigned to map the Unknown Landscape beyond Galala Valley! And Margo is requested to be in the party! He can certainly go. It's time he put his training to good use!"

"I'll go, of course!" Margo stated after seeing the parchment for himself. Myna took the time to inspect the parchment also. Samson looked at her with request.

"He's thirteen, and able to make up his own mind! He has my consent as well." Myna replied.

"You'll have two Dwelven scribes to map the land... one of them is Stewart Wildlife, Sassy's Dwelven uncle."

"Danger at every turn." Margo suggested. "Going into unknown regions, who knows what danger awaits! I'm in!"

"Let's visit tonight and you can leave come dawn. Margo needs time to pack his supplies."

At that Margo went to work on that very thing. Samson offered his visitors some ale. Then he excused himself, saying that he had to finish shoeing the horse. He would not be long.

Dawn broke on a new day and the troops were packed to the hilt. They had a long ways to go. The trip would take six days. Thonolan led in the front with Margo and Myna in the coach. It took til dusk that day to come to the Dwarven Terrs' border. There they slept the night. They woke up before dawn and headed west across the Principality. A straight line would take less time!

Four days passed, as the troops traveled west through grassy plainsland, which held wild food. On the fourth day the troops noticed a herd of wild horses! Margo stated that he remembered his dad stating that they were here, at this same place, during the Wildlife Campaign. However, Margo had never been here himself.

Passing the herd, they approached the Elven Forest from the north entrance. Thonolan noticed thick brush hiding the trail. He dismounted and led his horse and the whole troop through the bushes to appear at the northern end of the Elven Trail. He remounted and decided they would continue traveling.

They passed three villages as the villagers smiled and waved. Three hours later they came to a bridge which forged through the Elven Terrs, connecting north and south. Passing over, Thonolan knew that the trip was almost over... for now! Village Wildlife was the closest village on the south side of the bridge.

Thonolan waved the way into town.

The villagers cheered the arrivals. Ludwig Wildlife heard the cheers and appeared. The cheers also attracted Anna Littleton, who was aged at 15. She was instantly at Margo's side.

"You ready to blast the Unknown Landscape?" Margo opened.

"I'm looking forward to it!" Anna assured.

"When do we depart?"

"The troop's almost ready. Stewart is coming with a fellow scribe and they will be in charge of drafting the landscape. I just cannot wait!"

"I feel the same way!"

"We leave tomorrow! The sooner we get through Galala Valley, the closer we get to newness. I heard a rumor that Centaurs, herds of them, may live beyond it."

Dusk had come. The villagers had calmed and the visitors were led to shelter. Ludwig sat close to them to speak. He had aged nicely. Putting aside his walking cane, he had a habit of swearing every time his arthritis attacked. Everyone sat in a circle. Ludwig spoke.

"I am an old man! 78 years, to be exact! I'm too old for venturing. But you are all youngsters. You know the land and have learned to live on it. You will be given a treasure of supplies... but even that will run out, eventually. I am giving you one year to map the Unknown Landscape! It would take too long to come back for supplies! You will, at some period, be forced to live off the land. You have your weapons... put them to good use."

"Stewart Wildlife will be your guide and will be scribing the maps of said region. Gabrielle discovered Galala Valley, but you all are going beyond that, to the Unknown Regions. Working together, you will succeed. You leave come morn. Go get some rest now. That is all!"

Thor, Margo, Anna and Thonolan all exited and headed to a last night's sleep.

First, Galala Valley!

Margo awoke to the rising sun... he realized that he had slept in. He got up and exited the Elven hut and stretched. Thonolan saw him and motioned him over.

"We're ready... your riding horses are saddled. But first, the introductions." Thonolan led him to a crowd of members and family saying their goodbyes.

"Stewart Wildlife, this is Prince Margo Trellis of the Dwarven Terrs."

Stewart Wildlife, being Dwelven, was 5'5" tall compared to Margo's 4'9". Margo noticed that Stewart spent a long time talking to Sassy, as he knew that they were niece and uncle.

"Welcome, Margo! We're going to have lots of time to get acquainted."

Roderick Bloodbath, a previous member of the Avengers, pushed his way into the crowd.

"I'm getting a little bit edgy." He confessed.

"Roderick, you know Prince Margo."

"Certainly." Margo noticed Sassy in the distance saying her goodbyes to Anna!

Soon, seven horses, saddled and packed, were ready. The group finally split up!

The adventurers were made up of seven members: Margo Trellis, Anna Littleton, Roderick Bloodbath, Stewart Wildlife, Thor Tralesta, Populas who was a cleric, and Samuel, a fellow scribe of Stewart's knowing.

All of them saddled up going north towards Galala Valley. This was their first destination. The bridge was fairly narrow, so they were forced to go over single file. On the other side they picked up the pace and two hours later came to the border to the Elven Forest.

Going through the scrub they appeared on plainsland. The wild horses were laying, standing or eating the green grass. The troops watched for some time and then continued, following the Black Mountains north.

Two days' traveling in plainsland, following the Black Mountains north, brought them to The Gate, an open space where those mountains ended. It was thought by those who didn't know about The Gate that the Black Mountains connected directly to the Calicos, seriously blocking the people's passage. But the Elves knew better... because of the survey of a little-known Elven maiden named Gabrielle. She was able to predict the future in her dreams, it was said. Furthermore, Gabrielle acted on the dreams which led to her finding and surveying Galala Valley. Ever since the Elves had protected the valley and kept The Gate secret because of the special occupants within.

Palapany was a Centaur, part horse, part horse back attached to a human chest and head. The Elves protected her and Galala Valley in return for access to and out of the valley.

The troops had returned once again! But this time they would explore beyond the Gabrielle Mountains which stood on the west border of Galala Valley and which were named after Gabrielle herself.

The troops stood on the edge of the Black Mountains which ended here. Looking north you could not see the Calico Mountains which had been left far behind, to north and west was open prairie. The troops decided to go through The Gate into the valley.

During their travels the troops had watched for edibles which grew wild in the prairies. Wild carrots, wild berries, and edible leaves of lettuce, freely growing wild. Anna took special care not to miss anything edible... should stock up ahead of time, she notioned.

Palapany The Centaur

The troops had made camp after scaling the northern valley for two days. It was a clear night with soft winds. Anna was busy getting the evening meal prepared. But she suddenly heard thunder in the distance.

Looking up, she noticed that the sky was completely clear. What was going on, she wondered. The volume of the thunder increased... something was coming toward camp at drastic speed! She stood confused and speechless.

"What's going on?" she asked the others. Suddenly she saw a vision in the distance... a cameo colored horse was running toward them... its hooves hit the soil so hard that it sounded somewhat like thunder claps.

Everyone took notice as the horse approached the camp fire... it suddenly changed shape, and its humanoid form appeared. Palapany of Galala Valley was in their midst.

"I welcome you all to my valley." She spoke. "I know about your mission, and I concur with your plans! But I will also ask a favor. When you come back let me know if you see any others of my kind! I grow lonely in my old age. Soon I will die giving birth... but not quite yet, I may add. Let me know if I'm the only one... or not. Good travels to you." With that Palapany left their presence with additional thundering hooves.

"That was so cool!" Anna stipulated. "You heard her! We have two missions now! Will it ever end?"

The next morning the adventurers packed up camp and decided to head north into no man's land. They knew and considered that the trip into the unknown landscape would take them two additional days just to pass the border. On the third day they surveyed the length of the border going west. Prairies were abundant as the Gabrielle Mountains could be seen faintly in the distance to the south. Looking west, the prairie continued as far as the eye could see! They decided to turn around and head north instead.

Serpendel

The travelers had ridden for five days and camped five nights heading due north in line with the Gabrielle Mountains. The prairie seemed to go on forever without any leeway! The supplies they carried were at their max. Stewart and Sam were careful to measure distance as well as landscape. On the sixth night they set up camp, as usual, and had their evening meal. It was dark by the time supper was over. They all turned in early.

In the morning Anna woke up to the rising sun. She washed up, using some of their small supply of water bags, and glanced up at the horizon. Something caught her eye. Was this a mirage? She woke the others and told them to look directly north. It was not a mirage, it was a northern mountain range. It was blurred by the distance they were from it, and it could only be seen in daylight. They all cheered the change from grassy plains.

After breakfast they enthusiastically continued north. It took them two more days, but the mountains were beginning to appear clearly on the sunny days!

The third day came... they arrived at the base of the mountain range in the mid afternoon. Sam and Stew immediately began estimating the scale and distance of the range.

"Hey guys, look at this!" Anna exclaimed during the survey. She had found a cavern in the mountain side. A hole big enough to live in! She wanted to take a closer look and she stated so.

The entire team went in the entrance together. After ten feet the cave became pitch black.

"Light a torch, will ya?" Anna suggested. Roderick went out and brought back a lit torch. Anna went deeper and then suddenly hit something that blocked her way. She called Roderick over.

In torch light they found an old man laying on a raised hay mattress. He looked sickly white and they could hear groaning. Roderick brought the torch closer. Anna noticed red stains on the side of the old man's cloak. She reached down and felt the cloak. Her hand came back leaking blood!

"This man is wounded." Anna cut through the cloak and looked at the damage. It looked like a clawing which opened his side. And also, it looked infected!

"Give me a bag of ale!" Stewart got it for her. She immediately poured ale onto the wound. They heard more groaning from the patient. She then went to the horses and got her first aid kit. She began stitching up the wound.

"I'm afraid we are going to be stuck here for quite some time!" she notioned. "Let's just hope he doesn't get a fever!"

Anna was very attentive to the old man... and as the days passed he seemed to take on some color.

Five days passed and she would now remove the stitches. She did it fast and easy!

On the sixth day the old man gained consciousness... and stared at the intruders! He tried to rise from his bed so Anna was careful to watch his balance!

When he was on his feet he looked at his chest... all that remained were the scars! The patient began talking... but could not be understood!

"Which language is he speaking?" Roderick asked. The old man continued to talk.

"Hold on! I know that language!" Samuel stated. "I think he's talking in Cleric!" To confirm this Sam spoke some words... the old man replied openly.

"He's been isolated for years, so his common is a little wanting. But he will try!"

"Did... you folks... help me?"

"We were in the area... actually, we thought this cavern was vacant! You live here?" The old man nodded.

"My name is Serpendel... I am a Sor... Sorcerer!"

"Do you live here alone?" The old man nodded.

"Ever since my wife died six years ago!"

"Why did you choose to live here?"

"Where else is there? I can't go to Calesta City... they'd slaughter me instantly!"

"Calesta City? You mean there's civilization in these parts?"

"If you can call it that! Calesta City is a private town... you have to be chaotic to live there! And beware the Thieves Guild... they don't like strangers!"

"Is there anything else you can tell us about the landscape?"

"All I know is the southern prairies. There may be more land on the other side of Inland Islet!"

"We're surveying this land... do you know anyone who did this kind of thing before?" Serpendel shook his head.

"It was never done... simply because of there being no financial gain! It's all gold and silver! Watch out for Calesta City. That town is deadly!"

"We'd like to stay with you a while longer. At least until you gain your strength back."

"I'd appreciate that! The company, I mean. Who are you people? Where did you come from?" Serpendel asked.

"We're travelers... we are on a mission to survey and map these regions."

"Where are you from?"

"We come from the Elven Territories which lie on the south side of the Calico Mountains. It took us close to three weeks to get this far!"

"Beyond the Calicos? Your names, please."

"I'm Anna of Village Littleton. This is Margo Trellis, prince of the Dwarven Territories which lie on the north east border of the Calicos."

"My name is Thor of Village Tralesta."

"Village Tralesta! Where have I heard that name before?"

"What do you mean?"

"Village Tralesta! I think there's mention of that place in my parchments." Serpendel went around his bed and returned with rolls of parchment.

"Tralesta... Tralesta... ah, here it is! The history says that a couple, traveling, came to this land and settled on the land that is now a city called Calesta! The name was changed for some reason or another. But the founders of Calesta City came from your home, Village Tralesta in the Elven Territories. It also designates the location of Village Tralesta! I knew I had heard that name before!"

"Can I look?" Thor asked.

"Of course!" Thor took a glance at the records. Village Tralesta was placed exactly where his home was.

"Calesta City was my ancestors' domain."

"Of course the city was safe at that time... until all the chaotics took power! But now that's a nursing ground for chaotics of all types."

"I guess we'll bypass the city."

"Go around. We have to."

"There's no other way."

"Avoid Calesta City at all cost!" Serpendel instructed.

Two weeks passed before the party even thought of moving on! But Anna had been thinking... and she talked to everyone about her idea. She brought it up to Serpendel one day in the middle of the second week.

"Why don't you come with us, Serpendel? You're a wizard."

"Sorcerer! There's a difference in rank!"

"Come with us! You can't stay here alone and die of boredom. The troop agrees... all of us! We want you in our team! We don't have a human magic user. You'd be a great help!"

"In that case, I will come. I'm healthy, except for a huge scar on my shin. I need a horse!"

"I'll buy you one."

"Where?" Serpendel asked.

"Well, I could try a Centaur."

"Way too fast and way too loud!"

"Then ride double seated. I'm sure we could trade seats."

"In that case, I'm in. I'll just get my parchment ready and pick up my magic book for once!" Anna laughed at that.

Major Weaponry

Two months had passed since the explorers had left the Territories behind, but they were still residing at Serpendel's. Today was blustery with strong winds. Margo approached Serpendel and asked a question of him.

"Serpendel, do you know of any woodland around these parts? I need to make an archery bow." The old man took his time to think a few minutes.

"There's a forest three days' travel from here! In the middle of nowhere in prairie land. There's fine wood to be had there!"

"Thanks for the help!"

"No prob." Margo rose and headed toward Anna. He told her what he was thinking... she also thought it was a good idea to design new weaponry. She decided they would head out to the east the next morn.

They saddled up come dawn, as Anna called Serpendel to sit in front of her and act as a guide. Taking the reins, he headed eastward. After the first day he turned his horse south and headed in that direction until dusk. The next morn they turned east again, and at midday the third day they arrived at the woodlands.

"It's crammed full of trees! We have to go around it!"

"Exactly. But there is some ideal material for making weapons!"

"Set camp here, I guess."

"Right!"

After setting up camp, Margo went deep into the forest to look for the ideal branches for making a bow. Then he carved off the bark and shaped it to form. Connecting twine to the bow made a string. This was tedious work, but Margo enjoyed it and was obviously patient. The day after that he made another bow identical to the first! He walked over to Anna and gave her the second bow for herself. She began practicing almost at once. The arrows Margo had designed were steel tipped. After a few false starts she began to get more accurate.

She suddenly heard scuffling in the forest, coming her way! She hid until she noticed a white tail doe run into the prairie. She prepared an arrow and aimed for the neck. The doe turned into the arrow's path... and collapsed almost at Anna's feet. She pulled out a knife, and slit the doe's neck to bleed it. The deer twitched once, then died. Margo had seen all Anna had done... he literally ran to her and appraised her work. Then they gutted the doe and began butchering.

"Real meat! Anna, you're a natural! This will last us for eons!" Their dwindling supplies were recovered in style!

Thieves Galore

Two more days had passed. Anna decided that they would continue to the east come dawn. Dawn did come, and the troops mounted on cue. They started off, leaving the woodland behind, following the prairies east.

Mainly they found grassland and wild vegetables growing here and there. But they had all the supplies they would need for some time, so they passed by most of what they saw.

Four more days passed. Come the fourth day they could see a herd of horses ridden by unknowns. Anna decided to be extra aware... she stopped her horse and dismounted. She wanted a closer look! The others dismounted and followed her lead. She watched the other band closely. There looked to be five riders. And she noticed that they all wore leather armor. She took them to be thieves of some kind. She must avoid them at all costs!

She mounted again the next day and tried to avoid them. But she could not avoid a confrontation in the long term! Another band of five were heading their way! It seemed she had no way out!

As the thieves approached from opposite directions, Anna gave in, leading her band directly into the chaos.

"What's this? Strangers in our midst!" one thief spoke up.

"Let us pass." Anna insisted.

"Not before paying a tally, I say." Another notioned.

"1000 gold'll get you passed." Then all ten thieves laughed in the troop's faces.

"I warn you, we are trained in warfare." Thor suggested. "Let us pass."

"What's this? A white thief?"

"White or black, no matter! Let us free!"

"Tally first!" the ten thieves laughed once more.

"1000 gold? Nobody has that much!"

"I do."

"I do."

"I do."

"I do." The thieves all laughed. Finally, Anna was fed up! She reached back and handled her bow!

"You want a fight, you'll have one!" Anna equipped her bow and aimed for the one whom she figured the boss would be. She took him down first! The thieves regrouped but Margo followed Anna's strategy and soon, all ten thieves lay with steel tipped arrows in their guts. Anna walked around the thieves' corpses and led on.

"You may win one battle but I feel there's a lot more where they came from." Anna turned her horse around and faced Serpendel.

"You know something I don't?"

"There's a chaotic thieves' guild nearby! Those corpses were most likely members of that guild!"

"What?"

"You won one battle, but you just started a war, most likely."

"Thieves galore! You can't avoid them! Muster almighty! Then we win the war!"

"The population is about 200. Can we take that many down?"

"We just may as well try!" Margo suggested.

"This is beyond my belief! Two hundred against eleven!"

"We can succeed. We'll burn them out!"

"The fortress is probably made of stone!"

"Lead us there! We'll find out in a hurry!"

"As you wish!" Serpendel took the lead, sitting in front of Margo."

Two days passed with no incidence. They approached the fortress.

"A bark wood fortress! This will be too easy! This is a job for fireball!" Anna dismounted.

"Serpendel." She called. They crouched over to the gates, which were closed. Both stood together and cast spells.

Two fireballs hit the wooden gates so hard that they creaked and fell inward. Then they sent two more fireballs into the walls to the north. As the fortress burned in fury by the dried out bark tree logs, the fortress was forcefully evacuated. Not a soul was left behind. Anna and Serpendel, the rest of the troops, watched the inferno in grand amazement! That problem was probably taken care of now!

A Sole Stranger

Somewhere, far from Galala Valley, a sole stranger stood at the gates, what was left of them, to what had been a chaotic thieves' guild, at one time not that long ago! He could still see scattered ash and dust within.

He took the time to inspect the gates. There were two blackened marks in the wood of the gates, a two foot by fifteen foot gate mark which, he deciphered, could only be made by a fireball spell.

He deciphered that this had been done by two magic users standing side by side. He took the time to walk into the ruins and search for corpses. He noticed that there were none. That means that the fortress had been efficiently evacuated. He then checked further... horse tracks had been left not long ago. He noticed eleven sets of hooves. One set headed west... the others headed east. The tracks showed horses heading east! He decided to search elsewhere!

The stranger headed west! Two days later he found the thieves' corpses. Ten in all. All taken down by archers!

He headed east again... he decided to go find the band of eleven strangers who had taken care of a population of 200 chaotic thieves. His reasons were his own.

Another two days passed. The stranger realized he was already four days' travel behind the party. But he had all the time in the world!

Anna approached Calesta City.

"Are we stopping here?"

"Just to get Serpendel his own ride." Margo grimaced at that.

"Are you sure about this?"

"Just stay close." The team dismounted and Anna led them forward. Going into the populated city, they stayed snuggling together close to one another. There were many beggers in the corners of the walks. They noticed many thieves in the walks and avoided as many as they could.

The crowd was so intense that they did not notice that they were being followed. Anna approached the nearest stables and asked the price for a riding horse. The manager demanded ten gold pieces. Anna cringed at the price. She spoke to the others but they could only come up with nine gold pieces. Suddenly someone stepped up from out of the crowd and handed over the last gold piece.

"On me!" he insisted. When Anna looked up, she saw an aged man, perhaps 59 years old, which stood at 5'9". He was obviously grayed but there was not a single wrinkle on his face. He was wearing a leather robe, and around his neck he wore a cross medallion.

"You're a Cleric!"

"Allow me to introduce myself! My name is Catamar the Cleric." Then he took Anna aside. "I saw your handiwork back at the thieves' guild. Very impressive! You're not safe here! I will help you out!"

Leading the horse, Serpendel led the way to the closest exit. But there were ten thieves blocking the way. Catamar took a moment to cast a spell... then he pushed his hand out. The thieves all went flying into the closest gate and landed unconscious on the dirt entrance. One who was untouched opened the gate to let them through. Catamar nodded a thank you as he stepped around the unconscious thieves.

Escape And Map Readings

After escaping from Calesta City the troops jumped on their horses and galloped off at top speed. Anna noticed Catamar taking the lead by a long shot! She wondered on this!

Catamar slowed down...Anna noticed that he was leading them due south! Toward the Calico Mounts.

Two hours of straining speed later Anna slowed her horse.

"Catamar! Slow down." He stopped on a dime. "My horse is tired! We need to slow down!"

"Misty's not tired... not one bit! She could even outrun all of you!"

"We noticed."

"Ah, she's not just any horse! She's special!"

"How?"

"I'll show you some day! But right now, we need to keep marching!"

"Where to?"

"My home!"

"You have a home?"

"Yep, my home's in the Calicos."

"No kidding!"

"You can hide out there until the heat is off. Or, you can run."

"Our mission is to survey these lands!"

"No kidding! Well, just don't go back to Calesta City, is all I have to say!"

"Thank you for getting us out!"

"I'm partial to lawfuls."

"None of us are that! We're all neutrals."

"I fancied that! Let's get going!"

Two more days' travel brought them to the North Calico Mounts. Catamar led the way west until he pointed to a second cave in the rock.

"Home!" Catamar unsaddled his horse and then put her in a fashioned crevice outside the cave.

"Misty works wonders!"

"She's hard to beat!"

"I know... I chose her myself! She actually comes from the north!"

"How far north?"

"You could go on forever if you don't know where to look! Northwest, beyond Inland Islet."

"Is that close to the mountain range?"

"You know about that? You are observant! They're called the Euta Mountains. On the south side there's Euta Forest!"

"These have names?" Anna wondered.

"The Elders call this the Euta Vicinity! Everything has something dealing with the word Euta. It's just been called that for so long! No one knows how it started!"

"Could you come with me?" Anna led Catamar to the scribes who were already busy with their maps.

"I have names for the locations."

"Understood." Catamar pointed to the western mounts.

"Euta Mountain Range." Then he pointed to the forest. "Euta Forest." Then he pointed to the fortresses. "Chaotic Thieves Guild and Calesta City! Another thing, there's a bridge crossing the river. Right here!" He pointed far to the east!

"The Euta River goes both directions from Inland Islet. After a distance east Euta River opens into open sea."

"I know about the ocean! We all do! This is our home!" Anna showed Catamar a map of the known lands at home.

"Galala Valley?" Catamar gestured. "Why is that named that?"

"Maybe we'll tell you, some day!"

Surprise Visit!

Two days passed before Anna thought of leaving. Catamar had decided to escort the troops on their travels! And on the side, he would be extra security!

Catamar called the team to mount. "Onward!" he called. The team headed east following the mountains. Then, five days later, distancing Calesta City behind them, they turned to the north. Catamar led the way most days, and knew the entire landscape, he told!

"I'm a good guide in my own land!" Three days later they reached the bridge.

"What kind of bridge is this?"

"It's a suspension bridge."

"We have to walk on that?"

"It's perfectly safe! Even for horses!"

"Well, we'll give it one try!"

"Or two!" Catamar teased. "You have to come back the same way!"

"Oh, God!"

"Let's go for it!"

"Alright!" They took the horses single file onto the rope bridge.

"How long is this bridge?"

"Quite long, actually!" They traveled for two hours before they all reached the other side perfectly safe!

"I'm glad we're off that bridge!"

"That is quite a trapse." Catamar confessed.

"Which way?" Catamar seemed unsure of himself. "Catamar?"

"I guess if you want to map the landscape, you have to go in all directions. Let's continue north!"

Two days of traveling north revealed nothing more than more prairie.

"I'm tired of prairie!" Anna confessed. Suddenly Catamar turned his horse west.

"Now are you happy?" Catamar joked.

"Only if you can promise something different."

"Oh, what you want is a little action."

"Wouldn't hurt!"

"Then you shall have it! Follow me!" Catamar kicked his horse in the shins and took off at top speed. Anna did the same, but could not keep up! She went as fast as her horse could go... to no avail!

"He's teasing me and insulting my horse!" she thought. "What is it that's different?" She could not grasp it.

Two more days passed as Anna complained and pouted about all the prairie. But she suddenly noticed mounts to the west.

"That can't be the Eutas, yet!"

"Not the Eutas, something else! Something challenging!"

"A test?"

"You might call it that."

One more day passed before they reached a stone cavern in the middle of the prairie!

"What's this?"

"You must enter."

"Or what?"

"It's not dangerous! It's just a change of environment."

"I'd say! Team, follow me!"

"As you wish!" Catamar offered. Everyone went in the entrance. Roderick lit a torch.

"This goes so deep!" Anna realized. "Let's go!" The team followed deep into the underground caverns. These caverns reached so deep that Anna soon realized that there was a dirt roof above her.

When she stopped to rest she felt the creeps! Then a cold hand touched her shoulder from behind! She reached to remove the hand from her body... but the hand was so cold it was frosty. This caused her to turn around and she realized that she was facing a statue that lived... it was apparently made of ice! Then she noticed more statues appearing! She was literally surrounded by living ice statues!

"My friends!" Catamar suddenly showed himself. "These are my counterparts! They're friends of mine!"

"How are you, Catamar?" one of the statues spoke.

"How's the seize fire holding up?"

"It's not! They shattered two more of us! Ice just cannot hold up to stone!"

"My sentiments!"

"Living statues at war?"

"Ice against stone!"

"Stone wins every time!"

"We have no power against them!"

"What will we do?"

"I wish Exceptor were here."

"Who?"

"Never mind!"

"There has to be something!"

"We negotiate!" Catamar suggested.

"Gun powder would work! Blast them all the shreds!"

"We negotiate! Me and my team will go to them!"

"We will?" Anna suggested in surprise. "I'm here to do a survey, not help the innocent!"

"Now, now. Calm down, the stone statues are reasonable to a limit."

"What limit? Killing fellow statues?"

"We can at least try!"

"Okay, now you're the boss!"

"Indeed! Follow me, and we'll finish your survey... eventually!"

"We do have a whole year!"

"Trust me!"

"Okay, I'll play your game!"

Playing Her Hand

Anna was laid down for bed... but she could not sleep! The stone under her held no comfort! Why is this floor so red, she wondered. What could it be made of? She lit a torch. Then she looked closely at her surroundings! The whole cave was made of the same stone... then she scraped a wall. The stone came to shatters in her hand. Strange, stone that sheds! Wait a minute... she took a handful and held it to the flames of the torch. It burned into nothing.

"This is sulfur! Sulfur burns! If the stone statues knew of this..." Anna knew the dangers of the cavern's secrets! The stone statues could melt the ice statues with their own landscape! Then she got an idea... this would have to be worked on immediately!

The troops were gone to negotiate with the stone statues! Catamar, in the lead, led the others in the direction of the residence of the stone statues. Two days passed before they found another cavern similar to the ice statues' landscape. Stone statues met them on the way and led them forward!

Catamar entered the stone cavern and approached a stone statue with some status!

"I am here to negotiate a settlement."

"We will not settle!" the stone statue replied. "The ice statues will be obliviated... or they will leave our vicinity! One or the other!"

"You mean the Euta Vicinity in general!"

"We do!"

"Can't you cooperate?" Anna asked.

"We will not negotiate!"

"Well, maybe this will change your mind!" Anna knelt and tossed something into the watching crowd. It hissed and then utterly exploded, shattering tens of stone statues! "I have more where that came from!"

"Anna, you shouldn't have done that!"

"Why not? They use brute strength against innocents... I'm using brute strength against brute strength! Now the odds are even!"

"I guess she has a point!" Catamar relented.

"You're serious!"

"I'll take down this whole cavern if I have to! And you with it! I have more where this came from!" She reached in her bag and pulled out two more sulfur bombs. "Want a rerun?"

"Well..."

"The ice statues are not going anywhere. There will be no more deaths... on either side!" The stone statues all grimaced, but they agreed to relent!

"Just don't ever let us see an ice statue on our territory! We will kill trespassers!"

"Agreed!" Anna offered.

"Then it's decided!" Catamar walked quietly out of the caverns, followed by the rest of the troops.

"You didn't have to do that!"

"Oh, yes, I did! Brute strength for brute strength! I just evened the odds!"

"By killing how many innocents?"

"They weren't innocents! The whole band of them were on the same decision. They're all guilty!"

"What do you know?" Catamar insulted. Then he grabbed his horse and mounted. He said nothing. So Anna didn't see any reason to respond.

To No Man's Land

Catamar turned his horse around and headed back east. Anna silently followed behind and to the side as the other troops followed. They silently headed back to the ice statues' cave. Catamar spent the whole two days silently. Anna liked the quietness of the trip.

When they arrived at the ice cavern they rested the horses and went into the cavern without uttering a single word! The ice statues gathered.

"The stone statues have agreed to a settlement! They will not step on your territory! But you must return the favor! You will not step on stone statue territory! That is the agreement!"

"Then we will obey the settlement!" the representative assured.

"There will be no more deaths on either side!" Catamar ordered.

"We will live by those standards! Indeed, we will!" the representative promised.

"Then it is settled!" Catamar walked out as the others silently followed. Anna caught up with him at the entrance.

"You didn't tell them about my part!" she admitted.

"Less they know the better."

"Understood."

"Saddle up!" Catamar suggested.

"Where to?"

"To a land you have never set foot on and few ever find!"

"Drop a clue!"

"No way!" Anna shrugged her shoulders.

"I relent! But it better be good!"

"It is."

"How far is it?"

"Pretty far! It's a two month trip!"

"Then let's go!" Anna saddled up and headed west as Catamar took the lead.

"We're further south!" Anna noticed two hours later.

"I'm avoiding the stone statues." Catamar explained.

"That makes sense."

"We follow the river. All the way!"

"Means for good water... we can water our horses that way!"

"Indeed."

The days passed and the prairies changed into tall grasslands. Anna picked vegetables in several places and assured a fresh supply. She knew the troops needed their nourishment.

The first month passed. Anna and her troops looked upon the Inland Islet... it looked like an inland ocean of its own. Catamar rode up to Anna looking at the endless body of water.

"From here it's no man's land! Not a soul here but us! And it will stay that way! All the way in!"

"In?"

"Just a term I use. Pay no attention to it!"

"Okay."

Hellhounds Or What?

The troops had traveled inland for five weeks. The grassland had given way to dusty plainsland once more!

"How much further will we go?" Anna wondered.

"All the way to the Euta Mountains, I think." Catamar suggested.

"What's there?"

"You'll see when we get there!"

"What is it you're hiding? It's really got to be something special!"

"Unbelievable!"

"Some kind of waterfall, perhaps?" Catamar zipped his lips.

"How much longer do we travel then?"

"Three weeks, I think. I did mention two months, remember!"

"Okay, I give!"

"Thank you, I appreciate your patience!" Anna nodded her understanding.

Another week passed. Anna had made camp for the night. Rolling out her bedding she relaxed and started carving a stick... she enjoyed this most of all as a favorite hobby to help her relax. Wherever she found wood she would carve!

"Still at it!" Catamar noticed.

"Yep." Was her only answer.

"Sack time. We leave at dawn!" Anna nodded and readied her bed. Two hours later she was awakened by heavy breathing. When she opened her eyes she stared into the nostrils of a reddish brown wolf. Then the wolf spat fire at her! She turned a full circle on the ground and yelled for help! This scared the wolf, and he ran.

Catamar was at her side instantly! She sat down to catch her breath. "What was that?" Catamar asked.

"A... a... a hellhound! Reddish brown wolf who breathes fire! I've never been attacked by one before!"

"There might be more where he came from!"

"Well, I don't know. They are usually stragglers. Loners! I think."

"I'll stand guard for the rest of the night. Try to get your z's." Anna rolled over and fell into slumber knowing Catamar was guarding them all.

The next morning Anna saddled up and kept one ear open for howls. She also watched the horizon, watching for signs of habitation. She was nervous after the attack, and kept eyes open for signs of hellhounds.

"I wish Samson were here right now!" Anna suggested.

"Samson?"

"Before he became royalty he was a Dwarven guide. He had his share of fights with hellhounds."

"He won?"

"Every time! Takes a lot of guts to tackle one of them."

"I can see why!" Catamar confessed. Serpendel nodded agreement.

Another week passed. Anna decided to move toward the river as they were still following the Inland Islet. Catamar followed quietly.

They entered a gouge in the plains which was situated at the very end of the islet. Further on Anna could see the end of the islet and the river in the middle of the break. The land continued south some distance. Then it curved into the river. White rapids bounced on the river in far sight. Anna set up camp as it was close to dusk. Then she started heating supper. The land was grass plainsland so she dug a hole full of dirt before she lit a flame. She cooked safely and well! She really knew her stuff. But she soon realized she needed to relieve herself. She asked Serpendel to watch the fire. She moved a distance into some shrubs. But she did not notice the tree trunk, hollow, sitting next to her. She knelt to relieve herself and then laid eyes on the she-hound that was nesting in the hollow trunk! Anna laid eyes on the female hellhound and then paused in her place.

Locking eyes on the hound, she finished her job... slowly she rose, not removing her eyes from the hound's, and then ran! The hound ran after her... but could not keep up! She slowed down and turned back to her nest of pups!

"There's, there's a female... hellhound nesting her pups! They scared me shitless!"

"More of them?" Catamar asked. Anna nodded. "We're getting out of here!"

"What about supper?" Anna asked.

"It can wait!" Serpendel doused the fire and then helped get going. The team headed north... out of hellhound country... they hoped.

Catamar later discussed the situation in deeper detail with Anna... from then on the land near the river was named Hellhound Prairie!

The Gate?

The final two weeks passed with no hellhounds seen. The troops had left the river behind and headed north to safer territory! Catamar instructed the troops to set up camp at dusk one day in the later days of their trek.

In the morning they packed up and headed further west. During the daylight sometime in the afternoon they were blocked off by a stone wall which reached 25 feet on all sides. It seemed to stretch north forever! But Catamar was relaxed and calm. He led the troops north, following the wall for two more days. The troops obeyed their commandant without question. On the third day they found an opening to the west! They could see a second wall heading north further on the horizon.

"Welcome to The Gate!" Catamar greeted. "We must go in!" He took the lead as Misty tapped her front hooves in strange behavior…then Catamar sent her running at full speed through the entrance. Her hooves moved so fast that it sounded like thunder claps.

Anna stood in silent surprise as Misty turned around… she changed into a Centaur! Misty turned back and raced ahead. Anna paced her horse to top speed… when she got through the entrance, her mouth dropped in complete surprise! There was an entire herd of Centaurs grazing on an endless pasture, it seemed. All were varied colors with shaded pelts which took her breath away! Catamar got to ground and held out his hands in glee!

"Welcome to Centaur Valley!" he greeted.

"They're all beautiful! Astounding, in fact!" Anna relented. "How many are there?"

"I've counted 200 myself."

"Centaur Valley?" One centaur, dressed in a golden fleece pelt, approached Anna and changed shape. She became a golden fleece back with gold chest and head of a human.

"Welcome to our valley. My name is Goldy! I am primate of Centaur Valley."

"How… Palapany was right! There are more of her kind! Many more!"

"Palapany?" the centaur seemed confused.

"She's one of you! She lives in Galala Valley!"

"Galala? We are associated with a Galala which left us to lead a new life."

"Her offspring is Palapany. That's the current resident of Galala Valley."

"I see." Catamar relented.

"Yeah!" Anna swooned. "Listen, all this time I was wondering why Misty could run so fast! I think there's one thing that Catamar forgot to tell us! It's that Misty is actually one of you!"

"I just thought I would wait til you figured it out yourselves." Catamar added.

"It's strange how I just couldn't put my finger on the connection. Of course, until now, Misty never went near thunder clap speed!"

"No need to! You would never keep up!"

"Now, if I were riding Palapany…"

"That would be a different story altogether."

"Yes, it would."

"Stay as long as you want! What more do you need? This, I think, is paradise!" Catamar suggested.

"We'll survey this valley and then consider our next step."

"Well, this is it! As far west as anyone can go! Finish your survey! Then I will consider the return trip!"

"As long as I'm in charge!" Anna insisted.

"Understood, commandant."

"You will follow behind me." Catamar accepted openly.

"We have four months to spare. Let's make use of our time, shall we?"

Survey And Recompense

Two days into the troops' stay in Centaur Valley, Anna awoke to sunshine, although the winds were blustering to the west. The troops had set up camp in the valley not far from the entrance. The survey had begun and the scribes were gone to do their thing! The primary members of the crowd were all situated together in one camp.

Anna rose from her bed and appraised her surroundings. Turning to see the whole view, she saw that Margo was standing away from camp, and she noticed that he was slouching. Catamar came to talk to her, but she just ignored his approach and walked toward her future fiancé.

As she approached she noticed him slouching more. Then she touched his shoulder and he turned. He held tears in his eyes.

"This is my fault!" she realized. "I've been avoiding you! And now you're feeling jealous!"

"Not jealous! Just a little bit lonely!"

"For me!" Margo nodded as he refused to let the tears drop. "I'm sorry."

"Catamar makes a great leader!" Margo relented.

"You're a prince. I'm your princess! From now on I'm yours!" Anna held her arms out and Margo relented. The closeness comforted them both... then Anna offered a kiss. It was comforting, and all of Margo's uncertainties melted. Anna let go and walked up to Catamar.

"Margo needs me." She insisted. "We are promised to each other." Catamar took this as surprise and nodded his relent. He simply walked away. Anna came back to Margo and hugged him some more.

"From now on, you're my second in command!" she offered. "You are welcome with me up front."

"Catamar?"

"This is as far as we go! Now we consider the return trip! The scribes are surveying the distance to the Euta Mounts. After that we head home!"

"I don't miss it one bit!" Margo confessed. "Although I do miss my parents. I wish they could have come too!"

"I wonder how Samson's doing without my mother! It's quite a while since..."

"He visited last. I know."

"He's a busy man."

"But he always seems to have time for his family!"

"I would love to consider him my in-law." Margo smiled at that.

"We have all the time in the world!" Margo assured.

One week passed with no sign of the scribes. On the eighth day they were spotted coming from the north. The troops patiently waited and watched them approach. When they dismounted three hours later they smiled in confirmation. The survey was finally complete! Anna asked to see the results.

"The Euta Mounts continue 8 days' travel to the north. The distance to the mountain border is one day's ride. So what's our next step?"

"Does the prairie continue past the mounts?"

"Looks like it, but we never got to the end of the range! It goes even further than that!"

"We thought we ought to limit the survey to an even point!"

"Understood!" Anna confirmed.

"Maybe sometime we ought to survey the land between the Gabrielle and Euta Mountain ranges and to the extreme west!" Margo suggested. "The land continues in that direction."

"Okay, smarty, license us for another two years! It's been seven and a half months and we still have the return trip. It's a month and a half behind schedule."

"The Elves didn't tell us how far to go. All they wanted was a simple survey!"

"Well, we're behind schedule!"

"Then we'll take another six months and be over our time limit!" Margo suggested. "No one limited our travel time."

"People will begin to worry!"

"Let them. Far as they know we probably wouldn't make it back!"

"You have a point!"

"A good one!" Margo affirmed.

"Then we head back and retrace our route."

"I suggest we plan to leave as soon as we can."

"Don't want to worry them indefinitely."

"We leave on the morrow. You think?"

"Sooner the better. Oh, I'm going to miss this place!"

"Just wait until the Elves hear about it!"

Departure And Betrayal

This would be their last night in Centaur Valley. Anna and Margo were cuddled up together in front of the fire. The others were gathered also. But at dusk the Centaurs began changing behavior. They paced themselves close to the campers and began a soothing song in Centaur form. The sound was one of beautiful whistling. When the song was done Goldy stepped up.

"As Primate of Centaur Valley it is necessary for me to come with you to Galala Valley. Palapany should have her own kind around in her days of trial."

"She's nearing her time!" Anna relented.

"I must be there when it happens."

"It's going to be hard, losing her!"

"Her little one will need your love!"

"Palapany wanted us to find others! Like her! She wanted to make sure she wasn't the last... or the only!"

"Galala came from our midst and traveled to that valley... but it's isolated, this Galala Valley!"

"We welcome your company."

"Then it is done!" Goldy turned and looked in the direction of her member Centaurs. "Frankencense, step forward!" A brownish orange centaur came forward. "You will now be the Primate in my place, as I will not be returning. I will be residing in Galala Valley from now on. I will be known as the Primate of Galala Valley. And, of course, I will be there for Palapany's offspring. When the time is right. Until the morn, we bid you good night." Then the centaurs moved away and the fire glowed lower.

"Tomorrow's going to be a big day!" Anna suggested.

"Let's get some sleep so we can go at day break." Anna yawned and moved to her bed. Margo bid her good night and settled in for his own bed.

The next day started mild and frosty. After everyone finally saddled up the troops headed out of Centaur Valley. Goldy retained her horse form as Margo and Anna led the way. Goldy took up the rear.

"It'll be a long trek home." Margo noted.

"Let's just worry about today. I don't want to think ahead."

The travelers made good time and noted to avoid Hellhound Prairie altogether. After reaching the Inland Islet they moved to fresh water. Anna was relieved to be in familiar territory.

As the days passed the troops made good time. They were half way through the Islet, and Anna noted that they had, indeed, avoided Hellhound Prairie.

One month went by before they found themselves in Stone Statue Territory. But Anna felt uneasy. She wasn't sure what was wrong, but then she saw it... dust in the distance... and a battle which should not have happened... again! The entire party picked up speed as the Stone Statues warred on the Ice!

Stone statues pulled off power punches which shattered ice to bits. Little crystals of ice were dirtying the ground as Catamar ran into the battle. Anna was right behind with her sulfur bombs. She tossed the bombs which took tens of stone statues with each blast. Soon the ground was also littered with shattered rock. Anna managed to call a seize fire. Two stone statues remained from the battle. The ice statues were suddenly more prominent.

"Tell us what happened here." Anna insisted.

"One of the children lost their way and ended up here. It was only a child... and yet, the stone statues take offense. We could not let this slaughter go unavenged." Anna did some quick thinking.

"Well, there's no more threat, I basically cleaned them out. Anyway, maybe you ice statues ought to come home with us. Let those stone slabs be! I offer you a stone cave... in a different land. Why don't you leave this vicinity and come to a land of peace! Close to our home there is a cave. Come with us to our land."

The ice statues discussed the offer among themselves, stating that further talk was necessary.

"We will return to ice statue territory and consider this further!"

"We have a job ahead of us now!" Catamar complained as he fell in place.

"Relax, once we're across the bridge, you are free to go home to your cave in the rocks. Meanwhile, I have my own little mission to achieve."

Shared Secrets!

Anna saddled up and led the statues at a pace. The statues followed as a cloud. At dusk they were half way between stone and ice territory. The statues thanked Anna for her help and her offer with a gift... a necklace made of pure emerald.

"We statues are made of concentrated ice, inflexible and hard to crack! Except when it comes to stone statues!" The wise man, named Jed, laughed at his own joke. "We are hard knocks! We do not melt... we shatter!" He laughed again at his own wittiness.

Anna and Margo talked that evening about a new living arrangement for the ice statues.

"We could move them into Sam Littleton's old cave." Anna suggested.

"There's tunnels deep within, aren't there?"

"Direct connection to Village Littleton, too!"

"We'll have to pass this by the Elders."

"We must do that in private conference."

"Then it's agreed."

In the middle of the second day they arrived at the ice caverns. The statues led the way down. Anna accepted their company for one more night.

Anna noticed a lot of ice statues... some of them were children. She also noticed females. Jed kept her company.

"We have decided on the trip south! We will rest tonight and follow your men to the land you promised us on the morrow."

"You spared your children!"

"And our wives." As Anna looked further in, she saw more adults of varying heights. "The children must set the pace."

"Of course!" Anna understood. "Listen, I have good reason to help! This cave... it's a hazard! The walls would ignite at the slightest spark. You're living in a fire hazard! This is not a cave... it's a sulfur mine!"

"Sulfur?" Anna scraped the wall and chunks came out in her hand.

"This is a flammable powder... sulfur! Very dangerous! I'm surprised you chose this place!"

"Now I understand your motives! We were not taught this!"

"This is a sulfur mine. Very hazardous. Lucky the chaotic stone statues don't know this! They could bring this whole mine down on us. If they knew!"

"Then it is right that we leave!" Jed suggested.

"I can hardly wait until morn. But this hole will hold for one more night!"

"With good luck it will!" Jed left with this in mind as Anna laid down to sleep.

Jed sought Anna out first thing the next morning.

"We are evacuating the caverns in preparation for departure. And that means you."

Anna stepped out of the cave and as she came into the light of day she noticed that everyone of the ice statues held a lit torch. This move took her by surprise.

Jed showed himself and whispered into Anna's ear. Then he faced the sulfur entrance.

"In the name of us ice statues I command this cave to fall!" The torches were all cast into the caverns. Then Anna took the time to cast a fireball. She spoke.

"In the names of the Elven Territories and Galala Valley beyond, I hereby command this sulfur mine to collapse."

Anna released the fireball through the entrance of the cave. The following blasts lasted a whole two minutes as the viewers stepped back two steps. The dust was flying from the fall and collapse of the entrance. Sassy grinned with success!

In the end this experience sobered the people present, as they realized one by one where they had been living. Now they would go to a new home... a home totally made of stone!

One by one the ice statues turned from the mine and began walking to the east! The troops would follow from behind and they would keep the chosen formation for the entire trip home.

The ice statues actually set a pace which was acceptable to all involved! Which meant they wouldn't be dragging their heals forever.

Anna and Margo were close together at all times. Catamar owned up, staying far in the back with the Centaurs.

"One more week and we arrive at the bridge."

"Or sooner!" Margo suggested.

"How do we get all of them across?"

"Very carefully!" Margo suggested.

"Will it take their weight?"

"Ice is heavy, but not clumsy. Look it how far we've gotten already! In three weeks we've covered 4 to 5 weeks' territory. And maybe we'll even break the six week rule." Goldy changed shape and opened a conversation.

"You do realize your horse will get pregnant some day!"

"When she does, I'll raise her offspring!"

"Just keep Misty happy! That's all I ask!" Catamar nodded his head.

Anna called a stop for rest. It was running onto dark.

"Set camp!" She got off her horse and stretched her sore legs. Margo followed her. Then he began knocking his knees together in rhythm. Anna watched and then suddenly joined in! Then Margo changed the dance, adding crossed hands to knees , then back. The dance became fast with unfaulted rhythm. They got uproariously fast but finally fell back to the ground exhausted. Anna turned over and gave Margo an intense kiss. Margo responded kindly and then spoke.

"You remember when we learned that dance?" Anna nodded.

"At the first Allegiance Festival!"

"We spent hours learning that!"

"And goofing up, of course!" Margo laughed.

"Sonata Dilling taught us that dance! Fantastic girl!"

"Time for sleep!"

"Agreed!"

Another two weeks of travel brought them to the flowing channel with the suspension bridge. The stone statues called a halt. Jed came over.

"This is a suspension bridge!"

"I'm aware of that!"

"It will not hold our weight." Jed suggested.

"Have any other ideas on how to get you all across?" Anna countered.

"I do have one! We hover!"

"What?"

"Like this!" Suddenly the statue's feet tucked away and the statue floated on thin air!

"You'll hover over the bridge?"

"We don't need the bridge! You folks will, however."

"Then we'll walk while you folks hover!"

"I'll meet you on the other side." Jed smiled as if it was a challenge. This Anna had to watch. Jed spoke to his followers and one by one the ice statues tucked in their feet. Then, two by two, they began to hover off the edge of the water. They were slow and crowded, but Anna and her men followed over the bridge to keep pace. The ice statues had hidden powers that Anna had not realized existed!

Jed led the ice statues over the river's crevice. The crossing still lasted two hours. The statues set back on earth and extended their perched feet once more. Anna and the troops arrived on solid ground seconds later!

"Welcome to South Euta Vicinity!"

Anna advised to set up camp right where they stood. She immediately went to talk to Catamar.

"I have an idea on how to conquer Calesta City and return it to the Tralesta family!"

"We lay siege?"

"The ice statues are impossible to defeat! You need stone!"

"So we ask Jed to lay siege to Calasta City?"

"What do you think?"

"You already destroyed the thieves guild. You could probably do it yourself!"

"They will run at sight of the ice statues! They'll think of that invasion as a curse! They'll never come back. They'll disperse to different lands. We'll hire someone to run the town which will be renewed as Calesta City... home to lawfuls!"

"You're probably going too far, you know! Even if Clan Tralesta was the founder of Calesta City, it's questionable. We can't ask the ice statues to do this!"

"If they do an all-out siege on their own nobody will know we're behind it."

"They'll do differently if they know we're behind it! This has to be a secret mission that none of us can be involved in! Strictly ice statues!"

"Understood!" Jed heard the two talking and approached.

"What are you two speaking so softly about?" Anna looked at Jed and then away.

"I don't know... it's too much to ask!"

"What is? Talk to me!"

"There's a chaotic city a ways to the west called Calesta City! The city is totally infested with chaotics."

"So?"

"I was wondering if your group could capture the city for us."

"And exactly what right do you have to claim this Calesta City?"

"It was founded by Clan Tralesta from the Elven Territories! And now it's time we took it back!"

"If you take Calesta City for yourselves you might as well claim both North and South Euta Vicinities as part of your own continent! Let others settle here!"

"Our job was to survey the land, not lay claim to it!" Anna informed. "We never expected the land to be already inhabited. We thought it would all be wild land."

"It was wild... it would be, with all the chaotics around here!"

"Except for those chaotics, the land is barren."

"We'll have to take care of the scouts! That's your first destination!"

"Scare them back home!" Margo put in.

"The statues are my own people!" Jed put in. "We will discuss this situation and come to a decision together. Meanwhile, let's get supper going! The statues will take conference after supper!"

The next morning the ice statues and the troops separated paths. The troops would stay at Catamar's cave while the ice statues would start their march. Standing 100 strong in five rows of twenty, they marched like soldiers. They headed east in a straight line, headed directly for Calesta City. Four days' travel brought them in sight of the city walls. Closer in they could see five leather wearing thieves on horseback. But the ice statues hardly even looked. Their main target was the Calesta gates. They would not be deterred from their mission. The horsemen looked on some time in disbelief and then turned heads away. Soon all five were driving up dust in retreat!

Samuel Dilling, a chaotic thief, called for the gate. Calesta gate was open and he went in. He shouted a warning! No one heard him until he approached the master of the guild who was one who escaped the burned thieves' guild. And he knew not who had attacked the thieves' guild! No one had seen the attacker!

Now, Samuel stepped up to the Master of the guild and bowed his head.

"Rise, my friend! What frets you now?"

"I was standing guard... I left my post because we are being attacked!"

"Who would dare attack us? There is no one else in these lands! Surely you jest!"

"They are... are..." the soldier physically shook.

"Samuel, what is it?"

"Living Statues! Living Statues! All of them. 100, I estimate! We're done for!"

"Living Statues? Nonsense."

"I've heard of some myself." Someone spoke up. "They're made up of some kind of element. Ice statues, stone statues, I've even heard of Living Statues made of precious metals!"

"Samuel, who is this man who speaks?"

"I know not, sir!"

"Who are you?"

"My name is Catamar."

"Catamar... the Cleric who lives on the mount?"

"The same, sire!"

"What know you about these attackers?"

"They are made of ice... and they are unbreakable! Invincible!"

"Is there no protection from them?"

"None, my lord! I suggest we run!"

"Run?"

"As fast and as far as possible!"

"Leave the city unprotected?"

"This city is about to be captured and destroyed by living statues... and you want to fight?"

"We cannot retreat!"

"Then we all will surely die!" This brought up an uproar in the crowd.

"We leave, or we die! By the way, you may not know... these are the same soldiers that burned down your thieves guild..." This took the master by surprise.

"There is no other option?"

"None, my lord!" Suddenly they heard a sound of thunder as the ice statues began storming the gates open.

"It is true, my lord!"

"It is true!"

"Look beyond... they are Living Statues. We must preserve ourselves! Let them have the city!"

"Where will we go if we leave? This is our home! Has been for ages!"

"There is a land you don't know about beyond the southern border! Where the Calico Mountains stop. Beyond is vegetation and wildlife... and you can also find towns beyond the Calicos."

"Towns... civilization?"

"And lots of pockets to be picked!"

Thunder clapped as the splitting of the front gate was heard. The gates flew open and one by one the statues walked into the city. There was no resistance and nobody tried to attack.

Jed stepped up to the master and spoke.

"I, as the leader of the North Euta Ice Statues, demand your departure from this habitat! You will spread out and live elsewhere whereas this city is being fortified for the neutral Tralesta family from the Elven Territories! This city will be returned and reinstituted for the lawful Calesta family from whom you stole this city and from whom said city got its name! Therefore, we give you two hours before we begin to damage your living space... and if you refuse, we will not be afraid to kill!"

The thief master looked once, then twice at Jed and then a third time. His forehead was profusely sweating.

"For the time being we will move to the thieves' guild... the one you basically burned out!" Jed did not comment any further. That night Catamar headed back to his cave to join the human troops.

"The ice statues will have to hold the city... indefinitely!" Catamar suggested.

"So the city is secure?" Anna asked.

"Yes! Safe as apple pie!" Anna smiled.

"Don't you think they might have recognized you?" Anna asked.

"It was a concern... but when attacked there isn't much thought... just retreat! They weren't any harm."

"Not with living statues at your gullet, of course not!" Anna walked away. Catamar hit the sack and all remained silent.

One Thief Less

The next morning Anna awoke to some noise. It was Thor, pacing in her camp. He seemed torn between two things.

"Thor? What is it?" Thor cried tears softly. He turned to her.

"I have to stay!" he opened.

"Stay? What you mean?"

"I have to stay at Calesta City!"

"You'd leave the troops?"

"I'd like to keep the statues. They'd work as soldiers."

"You'd stay at Calesta City?"

"I've been struggling all night with this! I'm part Calesta... I'm a Tralesta! Legally, that city is part mine. I want to govern it! At least until some other Tralesta family can take over!"

"I can't stop you! You will be dearly missed."

"I've thought of what mom and dad will think!"

"You're old enough to make your own decisions. Then we'll see you to your new home. I wish you the best."

The human troops stepped into Calesta City in safety for the first time. Anna saw ice statues at their stations. Anna led the humans up to the pedestal that the master thief had held. Then Thor stepped up.

"My name is Thor of Village Tralesta. I humbly claim this city for my lost ancestors the Calesta family, who were slaughtered in a siege by Chaotic Thieves. By your ancestors! But I hold no hard feelings! I am classified as a master thief. And I am a neutral class thief! I can govern this city, as I see it as an inheritance and my home. The statues will be a permanent security system within these walls. Either you accept me as governor or you find homes elsewhere! There will be no bloodshed." There was silence from the crowd... then one person clapping... more joined in... then full fledged applause.

"You're descended from the Calestas?"

"The Calestas came from my village in the Elven Terrs. Village Tralesta."

"Then he is the one!"

"The rightful governor of Calesta!"

"You claim rights to this city?"

"I do."

"Then you're accepted! Ladies and gents, meet our new governor, Thor of Village Tralesta!" Thor stepped down and entered the crowd. There were calls and whistles of acceptance. This city was made of chaotic thieves... were they not? Yet, neutral class humans were welcome among them. And living statues!

Next, Thor went to the thieves' guild to see how things were going. First thing he noticed was there were living statues helping rebuild the guild. Thor found Jed in their midst.

"What's up here?"

"Since the thieves will be living here we just thought we'd repair what we broke."

"Supposedly broke, you notice!"

"The doors were intensely flared. We had to sand the doors."

"That's Anna's fireball! You should see it some time!"

"Toxic meltdown!" Thor laughed at the thought.

"Sorry."

The next day Anna decide the troops would continue their travels... less Thor and the Statues. The members of Anna's party each wished Thor the best and they were finally on their final trek toward home.

One day's travel passed in complete silence! As they reached Euta Forest Serpendel rode up to Anna and Margo.

"I'll be staying at my cave. That should be our next stop."

"You would be a great asset in the territories!" Anna suggested. He shook the negative.

"I should have died the day you came... I was ready."

"But now?"

"I'll live a little longer!"

"Okay, then we'll bring you home."

"Thank you."

They arrived at Serpendel's cave that night in the dark. Serpendel lit a torch and settled down. He reached behind the bed for his chest. Anna wasn't looking when Serpendel placed a roll of parchment into her hand.

"This you'll need, I think." Anna looked. The parchment was the lease to Calesta City owned by Village Calesta.

"You'll need that for proof! I don't have any use for it."

"We're going to miss you, Serpendel!"

"People come, people go... I'm just glad you gave me another few years!"

"We'll miss Catamar, too!"

"He's got his life!"

"He wanted me!" Anna put in.

"He was actually a show off!"

"He did one thing for us!"

"What's that?"

"Without him we wouldn't have found Centaur Valley."

"Or Hellhound Prairie!"

"Don't remind me, please!"

First thing the next morning Anna and her troops left Serpendel's presence. Walking, they left the Euta Mountain range heading south. Before they knew it they could see the Gabrielle Mounts and open space to the west. Galala still stood a far ways to the south!

Through three more days of travel Anna began to worry about how Thor's absence would be taken. Tralina and Thonolan... how would they react? She started to feel regret at leaving Thor behind, no matter how safe he would supposedly be. She decided to head directly home! The sooner they got home the better! After all, they were already two months behind schedule!

Anna's Enlisted

Two more weeks brought them into the centre of Galala Valley. Anna called a halt and ordered to set up camp. Goldy asked to talk to her.

"You should head to the Territories as soon as possible."

"I've been thinking that too!" Anna confessed. "I hope they don't think I've done wrong. We do have the survey intact. But if I went too far..."

"Thor's concerning you!"

"I don't know what to think! Did I do too much? Get involved too much?"

"You didn't expect habitat!"

"Far as we knew nobody lived there!"

"You did the right thing!"

"I hope so."

"This is where we separate!" Goldy informed. "I will now join Palapany. Best wishes on the rest of your journey." Goldy converted her form and stormed off in a blaze of thundering hooves.

Two more days brought them to The Gate. When they arrived they found a scout watching out for them.

"Holy snats, you're back!"

"Two months late!"

"That's okay! I've been coming every afternoon since you've been expected."

"You came from the forest?"

"I live in the forest village down the road."

"Lead on then."

Six hours later they entered the Elven Forest headed for home. The villagers of every town stormed with pride and welcome as the troops passed each one. But finally they dismantled at Village Tralesta. The word did spread that the surveyors were back! The first person to appear was Tralina. She noticed that she didn't see her boy among them. She folded her hands in regret.

"I expect my son found some kind of trouble along the way?"

"He stayed behind!"

"Stayed behind? Why?"

"He's inlisted himself as governor of Calesta City!"

"Of what?"

"Calesta City!"

"You mean... there's people out there?"

"The land beyond the valley is called the Euta Vicinity and it's inhabited! We must talk to the Elders at once!"

Ludwig showed up and scrutinized them!

"You all survived!" Anna laughed openly.

"Thanks to my skill in harvesting!"

"The food you left with would never have been enough to last! Congratulations, you actually survived a fourteen month survey of Euta Vicinity as you call it."

"Did you know it was inhabited?" Anna asked. Ludwig shook his head.

"I did not!"

"But you thought it could be possible!"

"An inkling of hope, it was."

"You want to provincize the Euta Vicinity? Make it part of the known lands?"

"To do that we'd have to open Galala Valley to the public! It's the only way in."

"We could go over the Calicos!"

"Not fordable. Not recommended, either!"

Two days later Anna stood in conference with the Elders of the Elven Territories.

"Anna Littleton, we bring before you the results of our conference. Seeing you found life beyond the mounts, we feel that you did quite a good job. You worked from your heart... helped Catamar's friends the Ice Statues... found Centaur Valley, the place where Galala came from... etcetra... etcetra... etcetra... this body has decided to enlist you... as a licensed travel guide for the Euta Vicinity!"

"What?"

"You will be doing what Samson Trellis did for twenty years as Samson Leader. You've proven success in leadership, so we are planning to use you as a travel guide in and out of Euta Vicinity. I already have your first job."

"Job?"

"It will pay 200 platinum pieces and take two months to achieve!"

"What will I be doing?"

"You will travel to Euta Vicinity and deliver your party directly to Calesta City."

"Let me guess... would the party be Thonolan and Tralina Tralesta?"

"They will be changing that to Calesta! They will reinstate Calesta City as Elven property. And then Thor will have his whole family start new. Mom and dad will be there, so he will not be alone."

"My oh my, just wait till Thor finds out we've gone back. He's gonna blow his top!"

Back To Euta Vicinity

Prologue

You can't start one book without having a conclusion. Beyond Galala Valley was a survey of unknown land lasting a full year. But it has no real conclusion. That is where this book comes in. It describes a continued story line and ends in a solid conclusion.

I felt unhappy that I concluded Beyond so soon. But as all writers know story lines sometimes continue from book to book. That is why I wrote Back To Euta Vicinity. I sincerely hope you enjoy this story line and the amazing conclusion. Sassy Wildfire

A Problematic Mission!

"My oh my, just wait till Thor finds out we've gone back. He's gonna blow his top."

"There's more."

"Continue."

"In order to increase work for you as a guide, we have decided to open Galala Valley to the public!"

"What?" This took Anna by complete surprise. "What about Palapany? You do know she's pregnant!"

"We have considered Palapany's situation and have decided to move her back to Centaur Valley where Galala came from! I'm sorry, but we need Galala open because of the increased population of the continent."

"Have you realized how secretive our hiding of The Gate and Galala Valley has been?"

"According to your survey Centaur Valley is enormous. And it's where Palapany's ancestor came from."

"So, how long does Palapany have, anyway?"

"The pregnancy will last another two years! That's more than enough time to return her to her old home."

"Does Palapany know about this?"

"She knows of everything before anyone else does! She's a Centaur! Of course she knows of her return to Centaur Valley! It's already been decided!"

"I'll need time to talk to her and see... I do realize that Palapany has no choice of her own."

"She has requested this of us! While you were gone, Palapany notioned that she did not want to be alone for this term of hers!"

"She did mention that to me... I had forgotten! She did request I search for others of her kind!"

"And you found them. Her health and the continued health of her unborn offspring is detrimental. When The Gate is opened to the public Palapany must not be in Galala Valley! She will also be happier among her own kind! That is part of your assignment as well! You, Tralina and Thonolan will bring Palapany with you when you go. You have two months to arrange things and be back in the Euta Vicinity. But note, Palapany must be beyond Galala Valley before the two months have expired. Understand?"

"Full heartedly!"

"You're sure you're ready and fully informed?"

"I am sure I understand where you're coming from. Then it's decided!"

"This meeting is adjourned. Good luck, Anna."

"God speed to you all! Thank you!"

Anna walked into the residence of her mother, Sasselia Wildlife. Ever since the death of her husband Alvin Littleton, Sassy had been living in Village Littleton. However she had recently been promoted to Head Sorceress as Amanda Wildlife, Sassy's mother, had died in old age some months ago! Anna still missed her grandmother, but knew that her own mother would make an astounding Sorceress. They both now lived in Village Wildlife. Walking into Sassy's house she saw that Margo had not left town. She greeted him with some surprise.

"You've heard?"

"Yep! Sassy told me everything!"

"So, Palapany has a long trip ahead of her."

"You've forgotten about Goldy!"

"No, I haven't, smart guy!"

"What do you need to do before we go back?"

"We?"

"Of course, we! You think I'm going to cut out and let you handle hell hound prairie on your own?" Anna smiled in gratitude.

"Don't you think you should let your parents know you're okay?" Anna suggested.

"Of course I do. That's why I came. I need to know when to come back so I can join the troops."

"I won't go back for at least two weeks."

"Then I will be here at that time. I miss my parents, but a job is a job. Lots of folks leave their homes for work purposes. Why, my own dad traveled for twenty years himself, without an actual permanent home."

"The famous and astounding Samson Leader! I can't believe it's been that long since Samson retired from it."

"My dad was the best travel guide on the continent!"

"He was the only!" Margo shrugged his shoulders in relent.

The next day the two teenagers went separate ways. Margo left for the Dwarven Territories while Anna headed for Galala Valley. Anna's trip would be solitude for her. She was travelling alone. As she passed over the bridge separating north from south she sang a song she had memorized from childhood. The day passed and before she knew it she had reached the end of the Elven Forest. She decided to set up camp as the sun went over the horizon.

The next day she packed her bags and moved on. She was not paying attention to where she was going as Snowy White, her horse, had travelled the distance so many times the horse could lead her. Anna sat on the horse with head down and suddenly she realized that she had arrived at The Gate. But when she looked up she saw two horses facing her. One was colored camion white and the other had a golden brown fleece! The air was suddenly distorted and Anna recognized Palapany and Goldy in their centaur forms. It seemed they had been waiting for her.

"Mercy sakes, the result of precognition." She swore under her breath. She walked up and approached offering a large bow.

"Greetings, my friends! I notice you've been waiting for me."

"We understand that you want to talk to us!"

"Then why come to The Gate? I could have met you in the valley."

"That would be wasted time!" Goldy opened. "Remember, we only have two months to move."

"Goldy, did you give me the impression that you were here to stay?"

"Indeed, I did!" Goldy admitted.

"Did you know that Palapany would be forced to move?"

"The times are changing... in not too long Palapany will die. And her offspring will bring in a new generation. Yes, I knew Palapany would have to move." Anna thought on that one very hard.

"Then... you came with us to guide Palapany home."

"But, not just that. I have always wanted to see where Galala had wondered to! Galala was the first in a line of Centaurs to live in Galala Valley. But now our home here must be moved. Times change, creatures die, and new creatures take over! It's just the way it works!"

"I need two weeks to prepare for the trip back to Euta Vicinity. That's also when Margo will be back."

"Two weeks is fine! We only need to wait and watch! We'll meet you here on the stated date!"

"Then it's settled."

"Anna, you must watch!"

"Watch what?"

"I see a grand war coming!" Palapany offered. "Grander than any before! Magicians, sorcerers, thieves both chaotic and neutral, clerics, scribes, magic users all on either side. The land will be ruffled, torn, shattered, burned, cracked, opened, torn apart. I have seen visions of strife, chaos, death, the poisoned, the stabbed, the beheaded! Beware and be advised!"

"It seems pretty serious."

"The worst battle of all time! Either side may win... good or evil. But be warned, it is coming, and will come."

"When?"

"Now! Never! Always! I do not see clearly., Just visions in my sleep. Go now. And be careful!

Margo had been riding hard during the day but taking the chance to rest at dusk! On the third day of his travels he approached the Dwarven Forest, the border to the territories he called home.

Riding through the path he noticed that the lights were dark at Milner Settlement. Ever since Alanah had become Queen Mother the settlement had just sat there, unused. He silently passed by.

Some hours later he approached Trellis Settlement in a slow march. He saw his dad in the smithy working. He moved in almost silently and howled almost in Samson's ear. Samson looked up and almost fainted.

"Margo!" Samson ran to him as Margo dismounted. "You made it back!"

"You think I wouldn't?"

"Come inside! Mom's going to be ecstatic." Margo marched into the house and yelled.

"Mom, I'm home!" Myna Trellis looked from the kitchen and almost broke out in tears. He had been gone for almost 15 months. "Come into the living room. I have a copy of the survey." Samson seemed very interested to know what was beyond Galala Valley.

"This area was already inhabited! This area is called the Euta Vicinity. Check it out for yourselves."

"There's even a city!"

"Calesta City! It was actually founded by the Tralesta family. By the way, Thor decided to stay and fortify Calesta City. But I'm afraid I can't stay long. Anna's been commissioned as a travel guide for Euta Vicinity. Also, Galala Valley is being opened to the public."

"Why? Where will Palapany live?"

"You're missing an important part of the survey. Centaur Valley, where Galala came from."

"So, Palapany will return to her roots, then."

"She's also pregnant!" This took Samson by surprise!

"After birth she will die!"

"They always do!"

"Palapany's been a mentor to me. She was primary in the fight during the Wildlife Campaign. How will I live without Centaurs close by?"

"You could always visit Centaur Valley!"

"It would take months to get there!"

"Six to eight months, actually. And that's just from The Gate."

"How long can you stay?" Myna asked.

"Only a few days... five to six, I guess! But no longer!"

"Anna has her job set up for her. Travel guides are the bravest people on the continent... and the most paid!" Samson laughed at his own guile. Margo just smirked.

Outstanding Inspiration

Anna was heading home from The Gate. A war like never has been... and never will be again. Palapany was kind of lean on the information part. When will the war start... and why? These are questions that need answers... right now!

When Anna arrived back at Village Wildlife she thought about her next step. She walked in on the council of elders and patiently waited to be noticed. You never rushed the elders, even in a state of emergency.

"Anna, you look like you have something to say."

"It's actually something to ask."

"Shoot!" The elder offered to listen.

"Palapany told me a war was coming. One like no other, past or future."

"Palapany's got a big mouth, it seems!"

"You don't believe her?"

"No, no, I believe her. There will be a war, like none we've seen before."

"So, what do you know about it?"

"Absolutely nothing!"

"Nothing? No information... no date or time... no reason?"

"Nata, I'm afraid!"

"Maybe I should ask Palapany. I did once... she just jabbered nonsense."

"It appears you don't understand Centaur Precognition."

"What?" Anna seemed just extremely confused. "We're talking about war here!" Anna glanced the room and her eyes set on Stewart Wildlife, her great uncle.

"I think I can help explain." He offered. "I've studied Centaurs a lot since I moved back to the Territories, and my conclusion is this! Centaurs know what's happened in the past... when and why. They know everything that's happening now... where and why. But the future is different. It's always modulating... centaurs have intense visions of the future... but unlike the past and present, they cannot predict when, where or why in the future. They don't have any more answers to offer us besides that some time in the future this will happen!" Anna finally thought she understood.

"Okay, so what you're telling me is she doesn't have any more information than we do."

"Exactly!" Anna sat back thinking. Wouldn't Palapany be like a witness... and this was her statement. Wait a minute... if she's a witness...

"Oh, great! Oh, just suffocate me and get it over with!"

"What?"

"Never mind. Inspiration popped. Could this war lead the troops into Galala Valley?" There were frowns all over their faces. "You're not opening Galala to the public, you're evacuating the valley to prepare it in case of war!"

"Do you still want the job?" the elder wondered.

"For 200 platinum pieces, you bet I do! That would be a life savings to me."

"Then it's settled.

"I promise, I'll keep both Palapany and Goldy safe. After all, they're our head witnesses, aren't they?

Anna headed back to Sassy's. When she arrived she found Ludwig Wildlife sitting with Sassy talking softly.

"Grandpa!" She crossed the room and hugged him in glee. Ludwig released her and she noticed his glum composure.

"You've heard of the war, huh?" Ludwig stood up in surprise.

"You know too?"

"Can't we do something? Anything?" Ludwig was about to shake his head no, but just then his head began to glow and he buckled up in intense pain.

[I am back!]

"Exceptor, what are you doing?"

[Ludwig, you are growing old!]

"I know that! I'm 92."

[I need another resting place. I can help. Palapany asked me to participate in the coming war! I need a new resting place... a platinum sword is what I want! Ludwig, relax and the pain will stop.] Ludwig did as he asked and the pain decreased. [You forgot the procedure... I have not been with you for eons!]

"So, you need a sword again?"

"A platinum sword... no other will work!"

"But there's not a single platinum sword to be had! Not even in the treasury." Sassy informed. Then Sassy's eyes glazed over as she thought. "How would you feel about the sword at the chapel?"

[That would be ideal!]

"We don't have it!" Ludwig suggested.

"But Samson does!" Sassy reminded. "We gave it to him at the first pledge festival."

"He's three days away!" Ludwig reminded.

[I can wait!] Exceptor suggested. [Quickly, bring me to my resting place!] The pain suddenly stopped. [Until the chosen time.] The light faded and Ludwig relaxed.

"Looks like we're going to the Dwarven Territories." Ludwig suggested. Sassy offered to ride with him. Anna relented the negative because of her own task of protecting the Centaurs.

The next day Ludwig and Sassy both rode on horseback headed for the Dwarven Territories. Anna watched them leave and then went back to her own business!

Recruits New And Old

Anna stood in front of the elders once more! She handed a sheet of all her recruits that she wanted for the return trip to the Euta Vicinity.

"Hm. These are all the people you brought on the survey... every one of them!"

"Well, we worked so well together, I thought we should use all of them again!"

"You will not need the scribes with you."

"At least give me uncle Stewart."

"I'm afraid that's not my decision. You'll have to ask him yourself! But otherwise, I'm willing. For all of them!" Anna giggled her delight.

Included were:

Margo Trellis

Anna Littleton

Roderick Bloodbath

Populas

Tralina Tralesta

Thonolan the Brave

Palapany

Goldy

Stewart Wildlife

Later in the day Anna contacted Stewart Wildlife and he agreed to go... with gumption.

Three days had passed and Ludwig approached Trellis Settlement. Samson was back in his old routine and heard the hooves before they even appeared. Setting down his work, he listened as they got closer. Ludwig's horse was in front, but he grinned fully when he realized Ludwig had brought Sassy with him! Ludwig swung off his horse like a spry young man and approached with a bow.

"Greetings, your majesty!"

"Greetings, friend headman. What brings you here?"

"The platinum sword, of course!"

"I have many of them. Choose one."

"No, those aren't adequate. I need the sword you were given at the first Pledge Festival."

"The Life Sword."

"Yes!"

"Then come with me." Samson led them into his house and headed for the bedroom. Myna noticed the visitors, but she seemed busy.

Samson moved his bed and reached through an apparently solid floor where the bed hand been. He pulled out a sword sheath. The sword he handed over was indeed the Life Sword of long ago! Margo noticed the visitors and stepped up. When he saw such a beautiful sword he reached for it! Grasping it he suddenly felt a glow go up his arms. It climbed his shoulders and finally reached his crown.

[Margo Trellis, you are the first to touch the Life Sword. You have been chosen as the Guardian of Exceptor. If you choose to accept this gift you will be given power beyond your own might. But beware... the mold I will make to your brain will be forever! Do you accept the meld with Exceptor and all his great power?] Myna came running in and saw the glow of light encasing her son's head. Samson also saw it... he took his wife to a separate room.

"Exceptor wants to meld with Margo. I think he should."

"Who is this Exceptor?"

"It's a spirit from the Underworld... it's totally harmless to the user. Although he will have some side effects which we need to consider."

"Like?"

"His personality might change a bit... but Exceptor would give him power beyond belief. If this war really is coming we're going to need all the help we can get."

"How big will this war really be?"

"Big enough to move Palapany out of Galala Valley to safety!"

"You mean... the war could reach the valley? Our valley? The one we conceived in?"

"Yes."

"Well, hell, why didn't you tell me? Yes, I accept Exceptor too."

[Then it is done. The final decision is Margo's and Margo's alone. Margo Trellis, will you accept the meld?] Margo laughed at the thought of owning his own platinum sword. A platinum sword, the highest currency in all the land.

"I heartily accept!"

[Then relax.] The light went up his arm again, up his shoulders, and headed for the crown. This happened without any pain at all.

[I am now synchronizing the Life Sword to your hand prints.] Margo's hand glowed, then dimmed. [Let no one touch the sword except the one who yields it!] Then Margo's head dimmed and he seemed totally relaxed. He tried a few jabs and swings to get used to the weight of the sword.

[I just wish you had been alive at the Wildlife Campaign. You adjusted in total relaxation. You didn't even feel as stitch of pain.]

"That's only because all my life I've known that spirits are real, and they're nothing to be afraid of! Especially a lawful spirit like you!"

"Exceptor..." Sassy asked, "when you brought me back to the upper world, you told me your birth name... would you repeat it?"

[I remind you that I asked you to never repeat that name... and so far you have held your pledge. I grant you my name... I was a human magic user... I lived in Anasthasia... and my birth name was Arthur Lenningham.]

"Why exactly have you returned?"

[The coming war is going to be horrific. The dead and dying will morn for death. The catastrophe in Galala Valley will be so great that even I would not be strong enough to help all the passing souls! This increased demand for my powers and Palapany's request have forced me to return.]

[My own primary mission is to ensure the injured on our side don't pass over... the dying enemy will all go to the Underworld... but none will ever pass!]

"That seems sad... all those misinformed fighters dying for nothing... just because they're chaotic."

[Chaotics thirst for blood... it is an addiction, when you kill often enough you begin to enjoy it... isn't that right, Sassy Wildfire?] Sassy jumped at the use of her old nickname from other years.

"I'm sorry you feel that way. Yes, at one time I did kill chaotics..."

[You killed the father of your own sister's son...]

"Why do you torment me so much?" Sassy had broke out in small tears.

[I want to prepare you for what's to come. You killed Andrew... yes, you did... but that does not mean you're guilty! Andrew deserved what he got... he was seduced into producing a child... a half breed at that... in that way he's no better than any of the mutants of Colina. I do not tell you this to hurt you... how long has it been since you picked up a sword?]

"Since... since the Avengers dismantled."

[And why did you let your great skill wither?]

"I became a mother... and wife!"

[Wife to two men, I recall! One was a chosen and the other was your own partner in combat! You should have chosen Samson first! His pain at your blindness hurt him.]

"We were friends."

[But it could have bloomed! But you shrugged him... because of the stupid stigma against mutants! The pledge of allegiance should never have been broken!] Samson stepped up then.

"Exceptor, I understand your concern. I also was blinded by the stigma! Who would I have to marry... there was none of my kind!"

[Exactly! This stigma has been on and on and on... and it never stopped! Even now, there are those who are against the holy men called Dwelves. Who were all forced to become scribes... just so they could hide their faces in shame... shame that they do not deserve.] Sassy was crying in sounding grief.

[Suchi made one horrible mistake! He disbanded the Avengers. You are all members and offspring of the Avengers family. If Suchi hadn't disbanded the policing force the group could have handled the thieves before they became pirates... and killed Anna's father. Alvin Littleton now rests in peace... and so does Serpendel.]

"Who?"

"Serpendel was a sorcerer we healed in the Euta Vicinity. We met him there. How did he die?" Margo asked.

[He died peacefully in sleep. Died of old age. We should all be able to die this way!]

"Why are you telling us all this?" Samson asked.

[Samson Leader, you once felt helpless over Alvin's death.]

"I felt it would come, yes!"

[But you stood still and let him die. It is not your fault, Samson. You once said that Samson Leader would be king!]

"Indeed, I did!"

[Anasthasius himself trained you! Do this for me... for me and for Ludwig! He never thought you should disband... he thought as I do! He thought disbanding The Avengers was a mistake... and so did I.]

"So what should we do? How can we change the past?"

[You can't! But there is one thing you can do!]

"And that is?"

[Ludwig's dying wish... that you all join again in a strong fighting group. Restore The Avengers troop and fight as this continent's policing force. You and your younglings after you should do what Ludwig has always wished! Bring back The Avengers.] Ludwig stepped up and spoke.

"I have been thinking all these things during my life! Yes! I have! And of all things, to disband the strongest and wisest fighting team that has ever crossed this continent... it's disappointing! And Suchi disbands because of something that has always been a family matter... it's not fair to any of us. And I know what Leo Anasthasius was thinking. Oh, we don't need a policing force anymore. The Dwarven Clan War was over... all Leo was thinking about was giving the strongest, most trained Dwarven fighter his freedom! Freedom to return and choose a wife... when his only wife should have been his fighting partner, Sassy Wildfire! How I love that name!" Ludwig was standing in anger and grief, tears coming out of his eyes. Sassy hugged him and he accepted her attention.

"Exceptor, we thank you for all your help!"

[Stop!] The sound was as loud as lightning.

"Yes?"

[You will do as I ask? Reband the policing force? The Avengers?] Samson thought of it, and asked,

"How long has it been since I killed a hellhound?"

Results

Once Margo was in bed the adults gathered for what was to be a serious council meeting.

"Why didn't you tell me Samson was in love with me?"

"Samson only realized it when you were in your deep sleep in Galala Valley."

"You remember all the times we would sit in the green grass under the shining moon and talk?" Samson asked.

"I'll never forget it! But I remember all the times you drank and bled your heart about your royalty! The first time, in front of the fire, I was astonished that you told me that!"

"Why?"

"Well, by all means, we were on an even keel. I was an Elven princess and you were a Dwarven prince. That brought me to thinking of the sealed Pledge Of Allegiance."

"So you were thinking romantically!"

"Every acolyte learns about our history as a course at the Holy Shrine. We also were told the tale of Gabrielle. It was part of our history!"

"So, are we going to regroup?"

"Some of us are too old!"

"Myna's coming with us! She can be our medicine woman."

"Seeing that Tralina is going to live in Calesta City, we will need her."

"Who will govern the Territories if we both leave?" That stood out as a good question... a very good one.

"Margo can't, we need his power!"

"How about Alanah?"

"Alanah Milner?"

"Well, my mom can't! She's already gone!" Samson noted.

"She's the Queen Mother!"

"She's only 52. She could rule for a good 20 years or more."

"Would you ever reclaim your throne, Samson?"

"Maybe... after Alanah passes on."

"And after we're gone?"

"The children will run the troop."

"They've already been trained! They're old enough now!"

"And will be even older when the time comes." Samson put in.

When the time came for Margo to return to the Elven Territories Sassy and Ludwig joined him on the three day trip home. Samson would leave the same day for Anasthasia to talk to Suchi Evil Killer who was now Leo Anasthasius' secretary.

After Samson arrived at Anasthasia he went directly to the palace. Walking heavy footed on his short legs he caught stares from everyone he passed. As he entered administration he walked in and slammed a hand on Suchi's desk! Suchi looked up... and grinned widely.

"Samson Trellis, what the hell are you doing here?"

"I'm here on business."

"Really?"

"I want to do business with you."

"I don't understand."

"I want to regroup The Avengers."

"What?" This took Suchi by surprise. "Why?"

"Suchi, there's a war coming that you're not expecting. No one is!"

"Why should someone start a war? And how?"

"We don't know that."

"Why are you so determined about this?"

"Suchi, you disbanded the group only because we might, note, might have lost one member. Ludwig's death wish is to have the Avengers regrouped. For heaven sakes, Suchi, the man's 92 years old!"

"What exactly does that have to do with me?"

"Ludwig said The Wildlife Campaign was just that... a campaign. He's certain that what happened there... everything... was an internal matter... a family matter... and you pulled the team apart over one death!"

"Sassy was a prominent member and an Elven princess. I couldn't risk her life a second time." Suchi responded.

"And what about Roderick? You had to resurrect him! He should have died in that dungeon! He was shot with magic missile! He died, Suchi!"

"I get your point."

"Suchi, there is going to be a war, and we need to find out when and why. As head of The Avengers, I testify that this will be our first assignment."

"How many people do you have?"

"I have Roderick, Sassy, Myna, Margo, Exceptooooor..." This last name he dragged vocally. This got Suchi's interest.

"Exceptor is in your party?"

"He's melded with Margo now."

"Exceptional! Well, this is something new! Tralina and Thonolan?"

"They're unavailable!"

"Too bad."

"Will you do this for me?"

"Done!" Suchi reached into his desk and handed Samson a key.

"This is the gate key for Avengers Estates." Samson recalled.

"You can all move in when you're ready. But you're funding yourselves, you should know. And I expect to get a rent bursary every month."

"How much?"

"For you, two platinum pieces." Samson just shrugged his head and walked out.

Stunning Attractions

Anna was the first person to see the troops returning from the Dwarven Territories. Anna smiled at Margo as he dismounted. But she saw a difference of some kind... he strode up and gleamed at her. She looked down and saw the sheathed sword. Then she looked again.

"Why, Margo, you look so handsome. And different in an unknown way." Margo slid the sword into his hand. "A platinum sword? Dwarves don't have much luck with those." Margo did a few slashes and jabs. Anna immediately recognized his intense skills. Then she looked at the sword again. She caught her breath. "That sword is enchanted!" she realized.

"How did you know?"

"Only an enchanted sword can give a Dwarf your kind of skill. Increased stamina, increased speed and certainly increased accuracy!"

[You are very smart! Shows true knowledge and training!]

"What's he call himself?"

"Exceptor!" This caught Anna off guard.

"The Exceptor? The one that's been dormant in grandpa all these years?"

"The same."

"Why'd he choose you? I knew Exceptor needed a sword, but I never thought he'd choose you. Well, when we make love I'll be doing it with him, too!" She giggled and blushed.

"Is that true?" Margo asked.

[I'm not going there!] Anna laughed.

"Well, everything's ready if you are. We could leave come morn."

"Then make it so!"

"Until morning, then." Anna walked away with a big grin on her face.

"You must increase my charisma too, Exceptor."

[Maybe I do!]

Anna, Margo, and the rest of the troops were gone long before Samson reached Village Wildlife. When he did, he went directly to Sassy. He approached her and handed something over. She opened her hand and saw the gate key! She opened her eyes wide and raised her hands high in a shout.

"You did it! He agreed?" Samson nodded. "Oh, I'm gonna slash and cut. Slash and cut."

"That's Sassy Wildfire talking!" Samson suggested.

"Oh, we're gonna have our old estate back!"

"Just wait til the gate guards hear us!"

"Let's tell everyone else!" Myna walked over and greeted her husband.

"How'd it go?" Sassy held out the key. Myna stared in absolute shock. "He agreed?"

"Yes, after charging me 2 platinum pieces a month!"

"Oh, that's easy to gather. Actually, that's not too bad for a whole estate!"

"And we have to fund ourselves!"

"We can share half half from our treasuries."

"And any other currency we find or gather. It's all ours!"

"I just can't wait!" Ludwig approached from a distance.

"You managed it, did you?" The key was held for him to see. "Congratulations, to all of you! Samson, you have made my dream come true! Thank you!"

"Well I think you should thank Exceptor more for spilling the beans!" Ludwig just shrugged.

Love Bells Galore

Anna's troops approached The Gate. Palapany and Goldy were both there waiting. Even with being pregnant, Palapany could still race with speed beyond normal means. The air changed and the Centaurs stood full.

Anna moved past them and led the way. Margo was right beside her.

"It'll go faster if we bypass the Euta Mounts."

"We might need Serpendel." Anna suggested.

"Serpendel's dead." Anna took this in with some disappointment but finally agreed.

On the third day of their travels the troops reached Galala Shrine. Anna and Margo decided they would spend the night inside. Going north they both slid through the apparently solid north wall. Inside was a roomy but not spacious room. They set up their needs and essentials and set up to rest for the night.

Margo woke up in the darkness of dead torches. He lit a torch and stabilized it. The light that gleamed soon woke Anna. She looked at him and he turned his head. Then he couldn't take his eyes off her! She was the woman chosen for him and he for Anna. Both were aged at 15 years of age. Both were to be mated to each other. Margo turned away, not wanting to intrude on her sleep. But she spoke.

"Margo?" He turned. "Please, make love to me!" He moved over, removed all his clothes, and lay silent beside her and then gestured for her to do the same.

"There's nothing here that can't stand the light." He quoted. When Anna was unclothed they looked at each other. Anna had full breasts plum and filled out nicely. Their size was medium for Elves and he thought she might fill out more yet. Margo, on the other hand, was slim but short. However, his abs and arms were the joints of hard worked muscles. His flesh was brownly tanned from working in the sun all day. He had started following his father's craft, and his father's before him. Margo's tool was solid and ready... and after so long of loving each other, growing up almost always together, they had loved each other forever.

Margo put his lips to hers... she responded with giggles... he moved to her ear... more giggles... then he moved to her breasts. He gently bit one and suddenly growled. Anna fell into uncontrollable laughter. That's when Margo felt her water. He was sure that it had been surprising enough. He mounted even and realized that his head was even with her breasts. Amazing, he thought.

"Okay, this might hurt." He informed. He began penetration. She gasped in pain which to her felt like a pinch. He moved further in. She yelped a little bit. Farther! She closed her eyes in pain and they began to water.

"Okay, okay," Margo nurtured, "I'm all the way in!" Margo pulled out and Anna began to calm down. He pushed in again watching his speed. Anna realized that now she felt bruised but it did not hurt. Margo was at a state of lust. His control had raised his need intensely. Soon he was flowing thoroughly in her passages and Anna discovered she liked it too.

Margo was at his peak... he stopped suddenly. He put his fingers down to water her... then he pushed in forcefully... Anna suddenly went into her first orgasm... Margo released at almost the same time. But with his love and true friend, the act had been ecstatic. Margo lay on her, satisfied and happy that this had finally happened between them. Anna, he noted, could actually get pregnant from this act, but Margo was willing to accept a family of his own... the Dwelven children of his own.

The next morning the troops continued further inland. Margo and Anna were all eyes for each other. Roderick Bloodbath, situated closer to the back, sped his horse and caught up to them. He observed them all smiles and laughter.

"I know what you guys did last night! It shows!"

"Does it really?"

"You two are made for each other, I can see."

"Roderick, you..." she was at a loss for words.

"Feels good when you love your partner, doesn't it? You two are hopelessly in love." Both of them had to agree.

"You'll be having your own batch of Dwelves eventually. Oh, by the way, don't let your kids become scribes, okay?" Anna and Margo both broke into hard laughter.

"The secret's out!" Anna suggested.

"If you ever want to do it again, feel welcome."

"Rod!"

"Oh, you're giving me a nick name now."

"Only if it suits you."

"I think it does."

Finally, Calesta

It took four months to reach Calesta City. When they did arrive they saw ice statues all over! The town was indeed still secure. Tralina and Thonolan both dismounted. Then they approached the gates. But two ice statues held them back. Anna ordered to let the couple pass. The parents of Thor walked a few paces... then they asked for help.

"We're looking for Thor Tralesta."

"Tralesta... you mean Thor Calesta, the master thief!" Tralina nodded. "Third door down." He advised. Tralina had never seen a living statue before.

They soon found the administration office. They took the dignity to knock.

"Come in!" They entered.

"How can I help..." Thor looked up and suddenly began to cry tears of happiness. Tralina looked her son over. He was now aged at 21 and he looked like a strong, immensely strong thief! He held 2 platinum daggers at his belt. But Thor had changed. When he had left th Territories he was still growing. Thor now stood at his father's height. He had his mom's nose and facial features but had his dad's forehead and eyes.

"You come so soon. I'm still arranging things. How long are you staying?"

"Forever!" Tralina informed.

"What?"

"That's right! We're staying here with you. You're gonna have to populate this area if it's going to be a city."

"That's already taken care of."

"What you mean?" Thor knocked on the back door. Suddenly a head popped out. It was female.

"Get out here, my mom and dad are here!" The mystery lady walked out... the girl's belly was bulged in pregnancy.

"It's yours?"

"Of course it's mine! This is Maddi. She's my secretary and wife."

"You got married?"

"Yeah, Catamar married us."

Other people began to file into the room. Margo and Anna, Roderick and Stewart, everyone was here. Thor looked them all over. He suddenly noticed Anna.

"Not you too!" The two women looked each other over, Anna and Maddi, they both had bulged stomachs in pregnancy. Thor looked a little concerned.

"You two tell your parents what you're doing?"

"Thor, don't but into our sex life!" was all Anna said. Thor just turned and walked around his desk.

"We heard about Serpendel. We're sorry!"

"He lived a fine life... he was old. I did make sure he got a proper burial."

"Is Catamar still at his cave in the bunedox?"

"No, Catamar runs a church in town now. He lives farther down the street. I'll lead you to him."

Thor led the way down the street until he reached a turn and right there on the corner sat The Church of Calesta Clerics. "It's funny, because Catamar is the soul member of this church. He's planning to get more clerics coming. From somewhere." He walked in and yelled into the distance. "Catamar, Catamar, you got visitors." Someone called down the hall... Anna recognized Catamar's footsteps anywhere.

"For heaven sakes, it's the troopers. Back already?"

"We were assigned to bring the Tralestas to Calesta City. We have a favor to ask of you. Bring them in." Anna shouted. Roderick brought in Goldy and Palapany. Catamar stood surprised to see Goldy back. But the other one must have been Palapany. She was not from Centaur Valley, he could see. He looked at the features and differences in Palapany. She suddenly changed to her Centaur form.

"Excuisite! And pregnant, yet. My goshes."

"We'd like you to return Goldy and Palapany to Centaur Valley."

"Both of them?"

"Yep!"

"The great Galala's offspring... and a newborn coming. I can't take them without repayment. Is there anything I can do in return?"

"Yes, there is, actually! We want you to come with us upon your return. We'll wait for you. As you can see, Anna's not in any shape to go any farther."

"How far gone is she?"

"She's at three and a half months and holding strong. Tralina says the child is healthy."

"You want me to come with you where?"

"To the Principality, of course."

"The Principality of the Crane? I don't understand why you want me there."

"We need your skills... desperately! Our parents are rejuvenating The Avengers. We need your service in our travels. It's actually a policing force which deals with Chaotics. But we do not have a single Cleric in our band."

"A band? A troop? Dealing with chaotics, undead creatures, etc, etc, etc?"

"Yes."

"Well, if you're going to put me to work, yes, I'll come."

"We'd be much obliged. By the way, we heard about Serpendel. We're sorry."

"Such a loss! A very strong sorcerer. One day he's there, then poof... he's gone to the other world."

"Do we have a deal?"

"We do! We sure do! Come ladies, we're off to Centaur Valley!"

"How long will you be?"

"Oh, about a month at most."

"One month?" everyone yelped.

"Sure, at top speed, a Centaur can make it there and back in about a month." With that he left their presence, the horse-centaurs following.

"Maybe we underestimated Centaur speed after all." Anna suggested. "How fast exactly is thunder speed?" Everyone shook their shoulders in lack of knowledge.

Looking Back

"What are we going to do for a month?"

"We're pretty well stuck here."

"We got Palapany out on time."

"Actually," Stewart offered, "we got out exactly three days before due date."

"That's reassuring."

"Actually, it hasn't been all that bad a life. I'm just glad Thonolan's safe and sound. It seems like he is full of luck."

"What you mean?"

"You want the whole story?"

"You know it?"

"I am a scribe, my dear! I miss nothing! I search around for new information. I have done biographies on all eight Avengers."

"There were eight of them?"

"Yep! In the Wildlife Campaign there were."

"We want to hear about Thonolan, though."

"I'm getting there. Thonolan met Tralina on the outskirts of Colina. Her father had been on a journey and he took his only daughter with him. But her father became deathly sick. Thonolan led them to Colina. Her father later died. Thonolan was a member of the Colina thieves' guild."

"They're well known as completely chaotics."

"Exactly! So, put a neutral thief with them, what happens?"

"They kill him!"

"Oh yes, they did try! But Suchi heard Tralina's yell of fear. They pretty well cleared out the chaotics... and Thonolan became one of them. An Avenger."

"Thonolan was an Avenger?"

"He certainly was! And so was Tralina. She trained under Sassy as an acolyte and gained her skills. But you have to remember, Tralina and Sasselia are counsins."

"So, if Tralina hadn't met Thonolan..."

"Thonolan would be dead today and Thor would never have been conceived, much less born. But the luck is, Thonolan has a family of his own, he's living in Calesta City, and he is entirely safe. He never need fear for his life again... meanwhile, the chaotics gather their strength and attack the Elven Guard."

"So the pirates came from Colina?"

"Every one of them."

"One of them killed my father." Anna confessed.

"But the killer's body was right next to Alvin's in the morgue. His death had been avenged."

"I never knew that."

"Meanwhile, we wait for another war to start, which we cannot prevent. Not without a when or a why."

"Sassy hated Andrew for what he had done to her... when they found him... and Sassy grabbed Andrew in a head lock and..." he illustrated a slash across the neck.

"She slit his throat?"

"Exactly!"

"Mom was quite the fighter."

"You don't cross her without consequences." Stewart stated.

The month was coming to a close. Stewart had been spending his time with the band illustrating and explaining the biographies of the original Avengers. Sassy's biography had been quoted last.

Two days later Catamar returned to Calesta City. The troops packed up the same day and started their return trip less Tralina and Thonolan.

Sudden Crisis

The troops arrived in the valley once more. Anna was 8 and a half months gone. The labor could start at any time... and Anna expected the labor to start early.

Anna and Margo were two days into Galala Valley when Anna turned to Margo and grabbed his hand.

"It's turning..." she informed. "we won't make it on time."

"We timed the conception too soon!" he relented.

"I'll be alright." She assured. "But this baby is not going to wait very long." Margo noted this comment. He moved back to the end of the troop and approached Roderick.

"Rod!"

"Yes?"

"Anna's not going to make it to the territories. You run ahead and find us a midwife!"

"Where will I find you?"

"You can meet us at the Galala Shrine." With that Roderick turned his horse and headed home with a heavy clop at top speed. Margo returned to the front.

"I've sent Roderick for a midwife. Can you make it to the shrine?" Anna nodded confirmation.

"I can wait another day or so, but it's possible she might want to come soon."

"Let's just hope we can make it to the shrine."

"We give birth at the same place we conceived."

Anna spent that night resting while watching her unborn child's movements carefully.

The next day they packed and headed further toward Galala Shrine. They finally arrived at dusk. Margo set up camp and the couple spent the night alone together in the chapel. Now all they could do is wait!

Roderick Bloodbath approached the Elven Forest. Not even slowing down, he ducked his head and smashed through the undergrowth at top speed! Not slowing down, he reached the path and headed for the closest village.

When he arrived he yelled for help. Dismounting, he ran for help. The first person to hear was an Elven maiden who seemed familiar for some reason.

"Roderick Bloodbath, what's the rush?" Roderick looked and realized who she was.

"Sam Littleton?"

"At your service."

"I need a midwife!"

"What?"

"Please, we need a midwife!"

"Who's..."

"Anna Littleton is expecting. She could go into labor at any time! If she hasn't already!" Sam stood shocked and confused.

"Please, help us! She can't do it alone!"

"Of course not!" Sam replied. Then she turned and asked to have her horse saddled. "We need god speed."

"Their child is Dwelven." Roderick informed.

"I did think of that." The horse was brought and Sam turned on heel and headed out of the village with Roderick right behind.

Anna was buckled up in pain from her first contractions. The pain eased... Margo held her hand in concern.

"She's almost ready." Anna suggested. "We don't have much longer." Margo put his arm around Anna's shoulder.

"We'll make it." He assured.

"You seem so certain."

"I am."

Two days later, at the fall of dusk, Anna began intense labor... but help had still not arrived.

"I'm not going to make it!" Anna was laying on a blanket, nude, and set up to see.

"You'll make it. I'll deliver if I have to. You just tell me how."

"It's true, I have the knowledge.,, but I hope we don't have to do this ourselves." Another contraction hit her. She cried out in pain. "They're getting closer!" Then Anna heard horses through the wall... two minutes later Sam Littleton passed through the wall. She came to Anna and nurtured her. Pain came again. Sam went to work. The next morning Anna gave birth to a healthy Dwelven daughter. Sam handed the baby to the mother. She was obviously Dwelven. Anna wondered how tall her baby girl would grow.

"I need a birth name." Sam suggested.

"Galala Littleton Trellis!" Anna insisted. Margo accepted the name freely.

"Where could they be?" Sassy wondered. "It's been too long since they left."

"Traveling takes time. They did say eight months."

"But they only have three days. I expected Anna to be back by now." Sassy and Samson were sunning in the open. Suddenly, they heard horse hooves. "Is that them? Oh please, let it be."

Anna was in the lead as they entered the village. Baby Galala was hoisted in a bag on Anna's back. She stepped down and gathered her newborn in her arms. She walked over and held Galala out for Sassy to see.

"Mom, say hi to your first grandchild. Galala Littleton Trellis!" Sassy almost didn't comprehend. Her baby was a mother... at sixteen? Sassy looked at the baby... it seemed healthy and normal... a normal Dwelven offspring! Her eyes watered and the she thought about the promise! Samson and Sassy had promised their children to each other long ago! Then she thought about how Anna and Margo had been raised together... she knew exactly how much Anna and Margo loved each other. She almost broke out in sobs. She was so happy for her daughter. Samson, surprised but not unexpecting, grabbed his son in a meld. He grabbed Margo tight. Tears also started... but Samson was a man... his eyes simply watered.

Going Home

Sassy's house was plum full, the whole family was celebrating the newest member of the family. Samson got up and dingled the gate key to Avengers Estates. "We have a mission... let's go home!"

"Anna, we want you in our party." Sassy revealed.

"Party?"

"We want you to be one of us... we want you to join The Avengers."

"Me, an Avenger?"

"You're just as strong as I am." Sassy assured. "In some ways you are even stronger."

"What about Galala?"

"Galala will be given a nanny."

"I can't leave her behind! I just can't!"

"You won't! Galala will grow safe in Avengers Estates. She will know Avengers Estates as her only home!"

"I'm going to be one myself!" Margo advised. "We always considered that you would be too!" Anna stood confused and uncertain. But then she cleared her itchy eyes and smiled.

"I accept!"

[I do too!] Exceptor revealed.

"You've been awfully quiet recently!"

[Indeed I have. I have been resting... meditating... but the future is so uncertain and scrambled... I have not been able to find the reason for the war.]

"So you've been working afterall!"

[I never stop.]

"Okay," Samson spoke, "as leader of the Avengers, I'm scheduling our move for tomorrow. So pack what you need tonight. We leave at dawn. Man, it feels good to be back in service. I really missed this!" Sassy laughed.

"Remember, Suchi Evil Killer was the boss... but now you are!" Samson noted that.

"Yeah, I am, aren't I?"

Dawn came with a glittering sun. Samson was seated on his steed.

"Oh, Red Coat, I missed you so much!" Sassy soothed her horse's head gently. Red Coat was a horse that she had always ridden as Sassy Wildfire. And now she was happy that Red Coat was still with her. She mounted last and they were off.

The trip would take three days and nights. They stopped occasionally to water their horses and for meals. It was an enjoyable trip which ended safely. Samson approached the Colina gates.

"Open for The Avengers!" Samson yelled. The gate guards looked down from their posts and then disappeared. Soon the head guard creaked through the gate. He approached silently and checked with his eyes. Samson Leader, Sassy Wildfire, and Myna Trellis were recognized instantly. "Your highness," the guard greeted, "we were not expecting you."

"From now on you will address me as my lord."

"But..."

"I am only an old man with a mission. We are what's left of The Avengers. But we do have some new recruits."

"The Avengers you say?"

"We have decided to regroup as The Avengers! It was a mistake for Suchi Evil Killer to disband us."

"I hear he works as Leo's secretary now."

"Indeed, he does."

"So you're starting with fresh recruits."

"We are The Avengers... We always have been..." The guard moved away and twirled his index finger for passage.

"Thank you, my friend."

"Welcome home, Samson." The guard offered as they moved on.

Avengers Estates was a mansion inside an extensive lot surrounded completely by stone walls. The gate was the only thing made of wood. The walls were made of shining crystal rock. They could not be breached... not even with a grap hook.

Samson opened the lock and The Avengers entered. Samson saw the stables exactly where they'd left them. However, they were completely barren. He dismounted and the troop stabled their horses. It seemed so nice to be back.

Avengers Are Back

The royal king of the Dwarven Territories has released his role to the Queen Mother to come home to Colina as Samson Leader once more. His mission is to lead a new troop of Avengers and protect the innocent! It seems The Avengers have been reborn. The Daily Scribe

Avengers Reborn

In an interview the previous leader of The Avengers, Suchi Evil Killer, approved and stated that the band of Avengers who have moved back to Colina are approved and sanctioned by Leo Anasthasius himself! Suchi also states that he had been informed and asked to allow their union before they actually agglomerated. It seems The Avengers will stay with us for quite some time to come. The Weekly Writing

Smart Councilors

The first moment he got, Samson took the time to visit Boar's Head Inn. As he entered, Peter Huntingham, the barkeep, swore under his breath.

"His highness himself!" He went around the bar and bowed deeply. "I greet you, your highness."

"Please, I go by My Lord Samson now."

"Of course, if it pleases you!"

"We have to talk." Samson gestured to the kitchen at the back of the bar. They sat themselves and Samson spoke.

"I have it on good order that there is going to be a war... a war as has never been and never will be again!"

"War? After this past twenty years of comfortable peace?"

"That's just the problem. We do not know when or why this war will start. This is our primary and first mission."

"Let me guess... you want me to keep my ears open."

"Any whispered words... anything suspicious at all..."

"I'll contact you! I certainly will!"

"This interview will not be repeated."

"Top secret. Understood."

"Now, pour me an ale."

"I'll get right to it! Welcome home, my lord."

The Avengers held their first council meeting the next day. Everyone took turns going up the levitating disk. As they were seated Samson called the meeting to order.

"To our first and primary mission. We will try and succeed in finding out when and why the war prophesied by Palapany The Centaur will be established. We do not have a timeline, reason, or any other details. Our first mission will be to find answers to these questions. I open the floor to discussion."

Roderick claimed the floor.

"I think we should check out the local thieves' guild. They may be chaotics, but I think we should keep a close eye on them! Maybe the answers will come from them." Everyone nodded agreement.

"Then it will be done! Good idea, Rod. Anyone else?" Catamar put in a claim.

"I could go to the local church and try there! Maybe someone there might have heard something."

"By the way, Catamar, you should actually register as a member! The church could be quite a help." Myna asked for the floor next.

"Palapany described this war as a continental war. She said it would even reach Galala Valley. In my opinion the only thing that would cause such a horrible war would be a calamity... or an assassination."

"Continue."

"Who's the most unprotected ruler? Who owns the most land? Not counting Galala Valley or Euta Vicinity?"

"The land is mostly owned by the Principality."

"Leo..." Samson thought. "He's the highest king of us all! He owns the most land. No, it's not possible!"

"Samson, don't let your love for Leo cloud your mind." Myna reminded. "Leo is the assassin's target!"

"Leo has fighters protecting him. It's not possible!" Myna quieted for some time. Then she spoke again.

"Roderick, what's the difference between a thief and a fighter?"

"A thief wears leather armor while a fighter wears metal armor."

"Is that all?"

"They also use different weapons."

"But... would a thief be able to wear metal armor?"

"Well, yes! I suppose so! But thieves aren't trained with the sword!"

"You're evading my question! Would a thief fit in metal armor?"

"In all truth, yes, he would."

"Thank you."

Samson stood in turmoil with the thought of Leo Anasthasius' assassination.

"He trained me. He nurtured me. If it weren't for him I wouldn't have such power. To think of Leo dying..." He stood his ground with concern on his face.

"What did Palapany say?" Anna spoke up. "There's something Palapany told me about when the war would start. Three words... now... never... and always!"

"Doesn't make sense."

"I've been thinking and I think I understand it now. Connect it to attributes of the future. Nothing's written in stone. The war could start now, or be in the process of starting. But if we stop it on time it'll never happen. At the same time we have to remember that warfare will always be there."

"So, the target's Leo, then."

[I will speak!] Exceptor notioned.

"You have the floor."

[The attackers will obviously be disguised thieves. During the Pirate Wars they attacked the Elven Guard. A program run by no other than Leo himself. But there's something you've all missed. The Dwarven Clan Wars! It is a perfect example of something that could be done on a massive scale. The thieves may copycat the clan wars on a much bigger and grander scale.]

"Then Leo's life is at stake."

[That is my conclusion.] Myna spoke out.

"Leo has a barren wife... it's either her or him. He has absolutely no descendants. If Leo dies there will be no heir."

"Then the fight starts for continental rule. A battle for the position of king. If the chaotics win..."

"Evil will fill the lands. The Principality will be a copycat of Calesta City!"

"This scares the shit out of me!"

[We will deal with this! Me and Margo. We will go to Anasthasia.]

"I'm going with you!" Samson suggested.

[As you wish!]

The Rescue

Leo Anasthasius lay sleeping in his tiny bed. It was pitch black. The torches were out. One of his fighters quietly creped in. He raised his sword to strike... he was beheaded before he could act. Leo awoke to the sound of breaking bone. He felt water on his face. Suddenly there was light... looking in the dim light he saw a personage in his room... but it was barely a man. He made out that the invader was Dwarven. Margo moved the torch closer. He pointed to Leo's hand. The water he felt was actually blood. Then

Margo pointed toward the floor. There were two parts on the floor... the body of the fighter and his separated head. Margo spoke.

"Your life is in mortal danger, your highness." Leo seemed confused. Margo removed the dead fighter's chest plate! Underneath was the fabric of a suit of leather armor.

"Thieves!" Leo connected.

"Exactly!"

"How many?" Leo asked.

"I don't know." Margo confessed. "I want your permission to kill all who attack us! I'm getting you out to safety!"

"Who are you?"

"My name is Margo Trellis."

"The Dwarven prince."

"The one and only." Margo offered.

"And I take it the Avengers sent you."

"Indeed, they did!"

"Get me the hell out of here!" Margo led the way reminding Leo to stay close. As they exited Leo's quarters he noticed the two slaughtered fighters... his room guards!

"They let the assassin in." Leo just grunted acceptance. They headed down the hall and a single being appeared at a steady pace. He appeared unarmed. He saw Leo and stood in shock. He saw the blood stains on Leo's chest.

"Are you hurt, master?" he asked.

"Not at all, Cheno. Not at all! Margo, this is my personal assistant, Cheno Seltin. I'm going to Avengers Estates in Colina. Be careful, this place is infested..." Margo could hear the marching fo fighters coming. When the fighters showed themselves Margo measured them up. Three against one. Possible but dangerous.

[We can take them. We can take them all!] Margo laughed at the possibility.

The fighters drew sword and attacked. Margo took out two of the fighters in the chest. Then he turned to the third. He was already down. There stood Samson behind the third with his platinum dagger leaking blood.

The troop of royals continued down the hallways toward the front door of the palace! Samson and Margo took out two more thieves on the way. When they approached the front entrance the fighters on guard let them through. Samson thought that rather odd. But as they stepped out they were ambushed. Thieves in metal armor surrounded them. They might be in trouble. But the fighters were stopped in their tracks. Leo looked at the fighters' stopped forms. This he found confusing. Then Catamar appeared and led them to safety.

"You paralyzed them!" Margo concluded. Catamar smiled.

"It's so nice to be needed."

Leo was moved to a safe house in the same town. Catamar and Sassy stood guard outside the front door.

"You all saved my life! I never thought they'd try this! An assassination!"

"We worked on evaluating the possibilities... you see, we just prevented a calamity." Samson said.

"Which would have led to war!" Leo hypothesized.

"I didn't want another Clan War!" Leo smiled. He knew what his death would have meant.

"It's lucky you rebanded... if not I would be dead."

"And there would have been a continental war as never has been and never would be again."

"I need an heir! Do I ever need an heir!" Samson smiled.

Epilogue

The palace was later scoured and both the dead and living thieves were disposed of. The thieves guild in Colina was stormed and defeated. But one thief might have gotten away... for evil is as much a part of life as good. Samson was sure that the chaotic thieves would always come back for more... that's where The Avengers came in! And one day, not too far away, a new Centaur would be born. As for the continent, it would continue in peace for at least another age.

In The Aftermath

Author's Notes

In this book, In The Aftermath, the Dark Ages or thirteenth century is coming to a close. My writing of this fantasy series spans the millennium from 1356 to 1394! These are only the recorded dates! However, the Pledge of Allegiance reaches as far back as 1300. The Avengers Trilogy occurs before 1356! The Dwarven Clan Wars occurred from 1336 to 1355. The pledge was reinstalled in 1326 and 1356! The millennium is ending... and so is this series! This will be the last of the fantasies by Sassy Wildfire! I truly hope you enjoyed reading this series, but please, keep looking for writings by Sassy Wildfire. Now, sit yourself down, and enjoy the concluding book of the Fantasies by Sassy Wildfire. My humble greetings, Sassy Wildfire January 23, 2009

Cast Of Characters

At Avengers Estates:
Samson Trellis/Samson Leader
Margo Trellis
Anna Littleton Trellis
Myna Milner Trellis
Galala Littleton Trellis
Exceptor/Arthur Huntingham
Roderick Bloodbath
Catamar The Cleric
Atamar The True
Sassy Wildfire/Sasselia Wildlife
Swina-nanny
Lord Eugene Brunweger
At Anasthasia:
Suchi Evil Killer – secretary to Leo
Leo Anasthasius – King of the Principality of the Crane
Samina Stranner Anasthasius – Leo's new wife
Baby Amana Anasthasius
Jacques
Mia
Chuck
Weena
China of Village Denvil
Adana of Village Denvil
At Euta Vicinity
Thonolan The Brave
Thor Calesta
Tralina Tralesta
At Elven Territories:
Sandra Derlin
Ana Derlin
Ludwig Wildlife
Sonar Milner
Samantha Littleton Milner
Joseph Littleton – grown son of Samantha
At Dwarven Territories:
Alanah Milner – Queen Mother
Satira Anders – Samson's head general
At Galala Valley:
Spirit the Centaur Offspring

Chapter 1

A nna Trellis, now the wife of Margo Trellis, went in search of Sassy Wildfire and Myna Trellis. When she found them both she asked them to attend. Searching for Galala Trellis, her 6 month old daughter, she picked her daughter up and groomed her outbreaking blonde hair. It looked like it would darken as she grew.

"As you know Galala is born a Dwelf. I don't want her to become a scribe like the rest. I want her to follow in my footsteps as a magic user. She could be a good addition to the troop."

"I agree!" Sassy informed. "When she's old enough to learn I'll take her as my acolyte."

"She's going to need protection for quite some time to come." Anna adored her daughter's short blonde hair. Suddenly one of the Colina gate guards came running in.

"I need to talk to Samson Trellis! Immediately!" Sassy went to find him.

"What's the problem?" Myna asked.

"There's a thief waiting outside the gates. He wants Samson immediately!"

"What for?" Anna and Myna both asked.

"The thief would not tell me anything. Not even his class."

"Chaotic?" Myna asked.

"Not likely. Probably neutral." Finally Samson showed.

"There's a thief at the gates that wants to talk to you. But you're not going alone!"

"Agreed!" The troop of Avengers which were available all left the Estates leaving Galala in safe hands.

When the troop approached the Colina gates Samson authorized the gates to be opened. The lone thief walked through and immediately approached Samson.

"Your highness, I greet you humbly!" The thief bowed deeply. "I bring word from Euta Vicinity. I am called Atamar the True. I am a lawful thief."

"Continue!"

"Thonolan the Brave asks you to accept me into your tribe. He called it a gift! Also, he says that he has asked Leo Anasthasius to consider Euta Vicinity as part of his worldly continent. With Thonolan owner of said land! Thonolan requests the title King of Thieves."

"Is there such a thing as a lawful thief?" Myna asked.

"As a lawful thief I am adequate with locks and dagger attack! The only thing I don't do is pick pockets."

"Come to our home!" Samson suggested. "We'll talk more there."

The trip back to Avengers Estates was silent. When they arrived Samson brought their guest directly to the conference room upstairs. Then he called a meeting to order. The whole troop was in attendance.

"I introduce Atamar the True. He says he is a lawful thief. And he requests membership on Thonolan's behalf. I open this meeting to discussion."

"Atamar, how did you feel about leaving Euta Vicinity?"

"Actually, when I learned that I would be searching for Samson Trellis, King of all the Dwarven Territories, I was most moved. He's been known as the most skilled Dwarven fighter in known history." Sassy spoke up.

"You folks do realize we have been without a thief since Thonolan left. We need him!"

"That's what Thonolan knew, also! That is why I was sent... he wished to have someone replace him!"

"Then let's take a vote... if any one of us says nai at all, then you must leave!" Samson warned. "All in favor of allowing Atamar into the troop raise your hands." The vote was unanimous.

"Then it's decided. I hereby call Atamar the True an Avenger! Welcome, my friend. Your help will indeed be needed."

Anasthasius In Love

Leo Anasthasius has put aside his barren queen... he now has fallen in love with Samina Stranner, daughter of Smithy Samuel Stranner. In a meeting with the press he has said that he will make Samina his new queen! And word is, Samina is in waiting with Leo's child!The Daily Scribe August 29, 1391

Old Queen Dies!

Leo Anasthasius' queen of twenty years has died in her sleep. Some rumors say it was suicide! Leo has asked for an autopsy. He says he will get to the bottom of things. He also says if it was suicide he will blame himself. However, he feels he had no choice than to put her aside. After the event six months ago when he was attacked he feels that what he did was necessary! However drastic the act! The Weekly Writing Sept. 1, 1391

Samson stood reading the latest news. Anasthasius had put his queen aside. Samson felt a visit was necessary. He would head for Anasthasia the next day. He also thought a journey would be good for his entire troop! He called a meeting that night.

"I now call this meeting to order! As you may or may not know, Leo has put his old queen aside. There are also rumors that his old queen may have taken her own life. I feel a visit is in order. So we will be going to Anasthasia come the morn. We will leave Galala in Swina's safe hands!"

"The quickest way would be by water!" Sassy suggested.

"The Elven Guard will make port in the morning. We'll get them to take us." There was agreement all around. "Then this meeting is adjourned."

Chapter 2

When the Avengers arrived at Colina Waterfront Atamar was the farthest behind. Going to board he saw a pickpocket reaching for a bag of coins in a stranger's pocket. He quietly snuck up and put the thief in a head lock. The thief struggled but as they watched the Avengers saw Atamar's intense strength.

"Give it back, you son of a bitch!" Atamar demanded. The thief struggled more, but to no effect. The stranger turned and saw the coins in the thief's hand. He grabbed the bag and struck the thief in the face! Atamar took this as time to let go. The thief ran in the opposite direction. The stranger reached in the bag and offered Atamar a platinum coin.

"Thanks!" Then Atamar boarded.

Two days of sailing brought them to Anasthasia Waterfront. Samson headed straight for the palace.

On entry the first person Samson sought out was Suchi Evil Killer. Leo Anasthasius' secretary.

"I thought you might come." Suchi greeted.

"How is he?" Samson asked.

"Sad for the queen but very happy to have Samina. Oh, by the way, the death was indeed suicide!" This brought Samson to feelings of dismay.

"Go in. He'll be glad of your company." Samson walked through a door into Leo's office. Leo turned to see his visitors... then he got up and gave Samson a huge hug!

"You're like a son to me!" he reminded.

"I'm sorry."

"Well, one must do what one must do!"

"You'll have a child of your own soon!"

"I just hope it's a boy!" Leo confessed.

"Is there anything we can do?" Leo shook his head. "There is one thing!"

"What's that?"

"There's a lady in town who says her attic is infested by monsters. I'll give you the address." Samson took it.

"We come to visit and you put us to work!" Leo smiled. "I'm sure we can handle it!"

The address was far into the residential area. Samson knocked on the front door which was opened by a lady with black hair and blue eyes.

"We're here to clean your attic!" She took them upstairs. Turning on the light she revealed two strange creatures skirting the floor.

"Carrion Crawlers! We must be careful!" The two octopus-like creatures seemed to sense their presence!

"I'll handle this!" Atamar assured. Pulling out a dagger he aimed carefully. The dagger dug deep into one of the crawlers. It wined once and then no more sound was heard.

The second approached. Atamar took it out the same way. When both were motionless he walked up and put both monsters into a pack sack tying it tightly.

"Garbage."

Samson noticed the dust and dirt spread around the attic. He went downstairs and spoke to the lady.

"Clean up the attic. Keep it clean and you won't have any more problems."

"Thank you!"

The troops left headed back to the palace. Leo smiled as Atamar put the sack of Carrion Crawlers on his desk.

"Carrion Crawlers!"

"That attic was filthy!" Samson informed.

"I'll dispose of these properly!" Leo assured. "How long you staying?"

"Until tomorrow. I felt we owed you a visit."

"All my old friends!"

"You got Thonolan's request?"

"Yes! Galala Valley will still be open to the public. And Euta Vicinity will indeed be considered part of the continent. I ordered new world maps to include Euta! You folks did me a favor by surveying Euta!"

"You can thank Anna and Margo for that."

"Yes, I can. Anna's got a child now!"

"Yes, a girl! She called her Galala!"

"And Prince Margo is the father!"

"True."

"Well, things, deep things will change as time goes by. After what happened I found I had no choice but to find someone new!"

"We must put our nation before ourselves!"

"That's the hardest part! Especially for me!" Leo admitted.

Chapter 3

The Avengers sailed back to Colina the same way they came. When they reached the Waterfront Samson led the way home.

Two days later Atamar searched out the Cleric, Catamar. Entering Catamar's quarters, he bowed down to him.

"Welcome, my son!" Atamar cried out in joy at Catamar's revelation.

"You know?"

"Ever since I saw you at the gates I knew."

"I am your son!"

"By Julie Andreak, I presume."

"Yes!"

"I loved her like no other woman! Ever!"

"What happened?"

"She was against me joining the church."

"So she was against you being a cleric?"

"Indeed, she was! She wanted me to herself! I could not stay!"

"You left her pregnant!"

"I am 58 years old. And as my son, you would be a young 28!" He stood up and embraced Atamar in a hug.

"Why did you choose to be a thief?"

"A lawful thief! As you are a lawful Cleric."

"That incident on Colina Waterfront may have earned you an enemy!"

"He never saw me! He was so busy with his escape!"

"The Carrion Crawlers impressed me. You handled them well!"

"They were easy targets! Too easy!"

"How'd you learn?"

"I was always infatuated with the tool. I taught myself!"

"You are very skilled!"

Catamar searched out Samson.

"I'd like to request a meeting."

"When?"

"Tonight. It would be considered a celebration. Could you arrange a pork roast?"

"Celebration?"

"Just be there! I'll explain then!" With this Catamar walked away. Samson stood confused for quite some time.

The troop was assembled for the meeting and pork was cooking across the room. Samson called the meeting to order.

"Catamar has asked us to convene. He now has the floor."

"Her name was Julie Andreak." Catamar began. "She was my wife!" But she was against me joining the church and becoming a cleric. So I left her to find my destiny... I left her with child!" Some of the troops were beginning to comprehend. Others wondered why he was bleeding his love life.

"Some months later Atamar was born! He is my son!"

"And I thought I was his only love light!" Anna joked.

"I say that if the Avengers had refused Atamar's membership I would have spoken for him."

Everyone began chatting in surprise so Samson quieted them down.

"I think their relationship should be kept hush hush! Only we will know! Time to eat! And I humbly say a full welcome to Atamar the True."

That same night Samson paid Sassy a visit. Her rooms were on the second floor. He walked into light and found that Sassy was still up. As he looked at her he noticed she was staring! It seemed to him that she had been looking at him a lot recently. Especially during the celebration, Sassy had eyes for him.

"How's it going?" He received no answer. He walked over and kissed her gently. This suddenly broke out into full fledged lust! She wouldn't let him go. He pulled away suddenly. "I'm sorry I haven't exactly been loving."

"I'm sorry about the pledge festival." She suddenly confessed. "You were right! Myna is your first!"

"No, you always were my first love. I was anguished when Sam kidnapped you!"

"Were you?"

"I kept wishing I had confessed my love." Sassy was sweating in the heat of the moment. Samson sat beside her and kissed her again. This time he could not pull away... he wouldn't even if he could.

Sassy took off her leather revealing her ripe, pointed breasts. Samson suckled for a moment and he was rising already. He hadn't been with her since the first pledge festival... and not even then. She removed her breaches... she stood nude in the centre of the second floor. Samson undressed.

Laying down sassy lay on her side. Samson faced her and his tool was solid! He kissed her and she laid down spreading for him. He laid on her and it naturally slid in. The entrance reminded him of other times... when the twins had been conceived... when he had cried himself to sleep during Sassy's coma.

He suddenly pushed in intense lust... Sassy had always made him exuberant. He pushed and pulled tightening his cleft. Sassy was wild with need... she could not get enough of him! He bit a nipple and began kissing down and down farther! Then he put two fingers in her cleft! She sighed in her need. Then he pushed even faster! She cried out and he released. She could feel his secretion as he gasped in delight. They just laid there beside each other all night long.

Chapter 4

The next morning Sassy sought up Myna.

"I want to talk to you!"

"Okay."

"You're a medicine woman!"

"Trained and true!"

"Well, Samson and I, we..."

"Slept together last night!"

"Yeah!"

"So what's your concern?"

"Well, I'm 43..."

"Spit it out, girl!"

"What if..."

"You're concerned about how I feel?" Myna asked.

"No, I know you understand."

"He's your husband too!"

"Well, I'm 43 years old!"

"So?"

"What if..."

"Are you ashamed?"

"No!"

"Then what?"

"Okay, I'll be brief! Can I still deliver?"

"Pregnancy? No concern!"

"Really?"

"I've delivered babies when the women were 45! No problem at all! 43! Man, you're young!" Sassy laughed her concern away.

"Actually it's been..."

"A long time, I know."

"Very long, in fact! I've been staring at Samson, feeling love and need and..."

"You know how to catch a man's eye! I never could do that!"
"Am I that attractive?"
"You're an Elf! Samson always did like your pointy ears!"
"But you have them too!"
"Dwarves don't have half the points of Elves!"
"I know I'm attractive."
"Alvin made a good choice of wife." Myna teased.
"Without him I wouldn't have Anna!"
"So how long has it been? Really?" Sassy thought.
"Since the twins were conceived!"
"I'll kick Samson's ass when I see him, I will!"
"I kind of hurt him at the pledge festival. All this stuff of being first!"
"You know, sex is what keeps a woman level."
"What you mean?"
"When a woman has sex it gives her a feeling of confidence. In herself!"
"Don't men feel the same way?"
"It's unfortunate that women only have so long!"
"Do they?"
"Yeah, when women get to a certain age we wither and dry up. That's when we stop producing offspring. Thank God we are both still young."
"How old?"
"Between the ages of 50 and 55. But even 45 year olds are safe if they have a midwife."
"So I have plenty of time!"
"And you're still fertile!"
"But..." Sassy could not complete her sentence as Myna had walked away. "Okay," she spoke to herself, "I'm good to go!"
As Myna walked by Samson she grabbed his arm and faced him.
"How come you denied your wife for so long?"
"What?"
"Sassy says she waited a very long time for you to notice her!"
"I know! But last night..."
"She caught your eye!"
"So what do I do? Tell me!"
"Treat her the way you should! She doesn't have all that much time left!"
"I don't understand what you mean!"
"She's just turned 43. If you want to continue loving her you'll have to make up your mind. She has maybe 7 years of fertility left. Samson, Sassy is yours. From now on Margo and Palapany will be my only offspring. Go to her, Samson. She's all yours! My mission, my only mission, is being a medicine woman. I have my mother's stature! I love you, Samson. Ever since I saw you on the floor with your blue green eyes I loved you. But Sassy needs you more. She always was your first choice!" Myna walked away in tears.

Chapter 5

Samson called a meeting to order.
"As you all know this year's Pledge Festival is coming up. We must all make preparations to leave Colina!"
"I'll not be going!"
"Myna?"
"I'm resigning my membership as an Avenger. Sassy needs you! I love you, Samson, I will until my dying day! But I've been chosen as medicine woman for Colina! I hereby release Samson Trellis as my husband!" This brought up an uproar. "I'm six years younger than Sassy! And maybe..." Myna broke out in tears. "Maybe I can find a fine gentleman..." Samson's eyes were watering.
"You're welcome to stay at the Estates if you want." He replied.
"I'll be available to take care of any injuries the Avengers need taken care of. But I..." Myna was crying fluently. "I can't stay and watch Samson's... Sassy, please, take care of him." Myna ran off the second floor.

"Samson..." Sassy was watching him carefully.

"It's alright!" Samson remembered something. "She chose me."

Myna Trellis Lets Go

The high queen of the Dwarven Territories has released Samson Leader as her partner in politics. She declares that from now on Samson and Sassy Wildfire aka Wildlife will govern the two territories as man and wife! And we have been told that Myna has decided to apply for the position of Medicine Woman for the town of Colina. I must say, she is sorely needed. The Daily Scribe September 23, 1391

Samson Trellis Heartbroken?

Samson Trellis, High King of the Dwarven Territories, aka Samson Leader, has told us that he feels heartbroken for the loss of Myna his queen, but he also says that he will renew his vows to Sasselia Wildlife at the upcoming Pledge Festival! The festival will be held in the Elven Territories on October 3-5, 1391. The Weekly Writing September 28, 1391

Sassy caught up to Samson.

"I love you, Samson. It hurts me to watch you hurt!" Samson brushed his fingers across her cheek.

"You were always my first!" Samson confessed.

"Remember the night you told me about your royalty? That night when we had the celebration? In Village Wildlife?"

"I'll never forget it!"

"That night I fell for you! You remember my tears?"

"Yes, I do!"

"Well, those were tears of hope and grief. Hope that I could somehow join with you and grief that it might never happen! That's when I fell in love with you!"

"So, you fell for me before I fell for you?"

"No, Samson! You fell for me that same night! It was me you talked to! It was me you told about your royalty! It was me you spent all those nights with... staring at the stars in complete celibacy! That was not missed!"

"So you do remember when we'd stay up late watching the stars!"

"And talking about nothing!"

"Maybe you're right! Maybe I did fall for you the same night!"

"The strange thing about love is... it crawls up to you and you don't realize it's hit you until later!"

"Like when I would sit up at night looking at the stars while you slept your deep sleep in Galala Valley!"

"Sometimes it's crisis that makes you think about might ofs!"

"So you chose me first!"

"You could say that!"

"I love you! Sassy Wildlife, I have always loved you!"

On the second of October the Avengers began their long trek to the Elven Territories. The route was so familiar to them that the distance was covered in two days! On October 3rd they arrived at Village Wildlife at dusk. Village Wildlife would be the host village for the festival.

Ludwig Wildlife, at 82 years of age, walked out to greet the arrivals. He walked with a cane.

"Samson, welcome! Welcome! Hello, daughter."

"Hi, daddy!"

"Well, the festival is started. Tomorrow will be your matrimony."

"Mommy!" Two twins ran to Sasselia.

"Hello, sweeties! How's the training going?"

"These are our twins? How old are you?"

"We're 12 years of age!"

"And we're living as acolytes! I always wanted to be a magic user! Not a scribe like Uncle Stewart."

"And if Anna has anything to say about it Galala will be an acolyte too!"

"Smashing!"

"Where is she?"

"She's too young to come!"

"Oh!"

Chapter 6

Samson spent that night with Sassy in his arms.

The next morning was a pancake breakfast and everyone was there. Except Myna! Samson was kind of glad that Myna had stayed behind.

Later in the day Samson spent time with the twins. They were always at Sassy and Samson's side. They were their parents, after all!

The matrimony started at dusk that day. Samson and Sassy stood holding hands. As Amanda Wildlife, the previous sorceress was dead, the sorceress for Village Littleton held the ceremony.

"We stand here today to renew the joining of Samson Trellis, Dwarven King and Sassy Wildlife, Elven Princess in holy matrimony. First, the lighting!" Sassy and Samson watched as a fireball was thrown into a fireplace full of drenched wood. All the same, the wood caught instantly.

Ludwig stepped up.

"Red ochre to enrich the fertility!"

"Circle the fire pit once, please!"

Samson and Sassy did as asked. Then Tralina Tralesta, who had come from Euta Vicinity, came up and threw in stones of charcoal.

"May your joining be like stone and never falter!" Samson and Sassy circled the pit again.

Samantha Littleton was last.

"The power of metal to keep your joining fast and hard as steel!" Sam threw in two platinum pieces. Then Sassy and Samson circled one last time.

"Stand forward!" the sorceress asked.

"Samson Trellis, as king of the Dwarven Territories do you take this woman as your wife in accordance to the scripture of the Pledge of Allegiance?"

"I do!"

"Princess Sasselia Wildlife, as new head sorceress of Village Wildlife, do you take this man as your husband in accordance to the scripture of the Pledge of Allegiance?" Sassy stood confused.

"I do!" This was almost a whisper.

"Then, by the power I possess by the government of the Elven Territories, I declare you man and wife!" There was applause loud as thunder. Samson and Sassy walked away befuddled! Ludwig met them on the outskirts.

"Well, congratulations!"

"What's this about..."

"Yep, it's happening!"

"Head sorceress?"

"Yep! From now on you're considered our head sorceress!"

"But I'm in Colina!"

"I know! I talked to the Elders and they all agreed to instate you as sorceress!"

"But what about Samantha?"

"She's been Wildlife's sorceress for quite some time now! Remember, her position was only temporary!"

"Then..."

"Yep, it's true! You're the sorceress now. Head sorceress, I may add!"

"But how can I work here if I'm in Colina?"

"We will keep Samantha on as local sorceress! It doesn't matter where you live! You're now Head Sorceress of all the Elven Territories!"

"I wasn't told!"

"I know you weren't! We made the decision and talked a lot about it! You're 43 years old and close to 50! You are prime meat for the position right now! Take it, it is quite an honor." Sassy stood stunned. Samson hugged her in glee.

"I'm head sorceress! I can't believe I've come so far!"

The third day, as always, was the day to pack up and enjoy leisure. Samson arranged to leave the Territories the next day!

Sassy approached her tent in small steps. She knew something was different in her own stature. Samson looked up and held out a hand. She took it and suddenly gagged, belching profusely. When it stopped she cleared her throat.

"Are you okay?" Samson asked. Sassy nodded.

"It's just morning sickness!"

"What?"

"That's right! I'm pregnant!" Samson was spellbound.

"We're going to be parents again?"

"I talked to Myna. She said that pregnancy would be alright for quite some time!"

"You're still fairly young! Why didn't you tell me?"

"Because I've produced more offspring than most!"

"Most of them Dwelves, I may say!" Anna Trellis approached.

"Mom, you okay?"

"Yeah, why?"

"Because you're pregnant!" Samson laughed out loud.

"You knew?"

"I do have a child of my own! And I heard you belching in mornings at Colina! I knew you had conceived again!"

"How do you feel about that?" Samson asked.

"It's alright. After all, you were both the first love!" With that Anna walked away.

"Did she listen to us, or what?"

"I think she's accepting it. She's got rocks in her brain! It appears she's heard a lot during her stay with us!"

Chapter 7

The Avengers arrived back in Colina two days after the festival.

The gates were opened and they stabled their horses. Samson stepped into the mansion... all was dead silent! Samson thought something might be wrong. The atmosphere seemed gloomy. He walked the main floor. Not a soul in sight. He checked the nursery. Galala's crib was barren.

"Myna? Swina? Anyone home?" He strode to the second floor. There he saw torch light. Sitting alone at the table was Lord Eugene Brunweger!

"Where is everybody? Galala's crib is empty." Eugene looked at Samson.

"Galala is fine. Swina and Myna took her to Swina's place."

"Myna?"

"Thieves broke in!"

"What?"

"I'm sorry you lost Myna!"

"It was her choice!"

"True!"

"My Lord, she chose me! When I got back I had no thought of marriage! I was just thinking about getting things done!"

"And Sassy?"

"I've loved her for a long time! She was my choice!"

"The thieves ransacked the building! If Myna had been here she would be dead right now!"

"I hate chaotic thieves just as much as you!" Sassy arrived.

"Galala's not in the nursery!"

"She's fine. Myna took her to Swina's!"Eugene spoke up.

"I believe the thieves were after Galala! They planned to hold her ransom. They have attacked us for the last time!"

Samson went to see Swina and Myna that same day. He told them to keep Galala at Swina's place for a while. Until he did what had to be done.

When Samson got back to the mansion he found a letter which the scribe had delivered.

Dear Avengers: We humbly request a battle to the death. Arrange a time at the coliseum. Yours truly, Colina Chaotic Thieves Guild

Samson thought of ripping up the letter but then thought better of it. He went directly to Lord Eugene.

"Those sons of bitches! They want a fight, they'll have one!" Lord Brunweger roared in rage! "I'll confirm the time and date! I'll take care of this!"

"I'm fighting with you!"

"You all are! I'll call in all our recruits. Thor, Tralina and Thonolan among them! We'll need every man and woman we can find! But not Sassy!"

"She is carrying!" Eugene nodded.

Battle To The Death

At the ransack of Avengers Estates the Chaotic Thieves Guild has invited the Avengers to a battle which will decide everyone's fate! Suchi Evil Killer and Leo Anasthasius himself will be in the battle. Leo has sanctified said battle and says the thieves couldn't possibly win! The Daily Scribe November 1, 1391

Samson awoke to Margo's face.

"We have to talk!" Margo insisted.

[Hello, my friend!]

"Exceptor! Where have you been?"

[This battle is important! You are right to accept their challenge! Revenge is sweet! This will end the conflict within our walls! I will be in the battle!]

"Take care of Margo!"

[I will keep him safe!]

"We need every man we can find."

[Samson, one word to cheer you up! Sassy carries a son!] Samson stood stunned.

"Am I hearing right?"

[I demand that Sassy stay at Swina's for the time being.]

"Done!" Margo sat down and hugged his father.

"I'm gonna have a brother!"

Chapter 8

Now in the coliseum. A battle as never before! The Avengers against the Colina Chaotic Thieves Guild!

Now the rules... there are none. Introductions. In The Avengers party... Leo Anasthasius, Suchi Evil killer... Samson Trellis... Margo Trellis... Anna Littleton Trellis... Tralina Tralesta... Thor Calesta... Thonolan the Brave. Catamar the Cleric... Atamar the True...

[He never called my name!]

"Spirits don't count!"

In the thieves guild... Master thief George Grayson... acolyte Peter Wenzy... acolyte Steven Leppy...

"Those guys look rough!"

[Just rough around the edges!]

Now the slanders. As king of the Principality Leo Anasthasius will speak for the Avengers.

"I'll say one thing... you ass holes are making a big mistake!" The thieves laughed. "Laugh now, because you won't later!"

Master George, your turn.

"You sons of bitches have given us problems ever since you were amalgamated. And now, this lawful thief of yours intrudes in our work! That was the last straw!"

Suddenly Atamar looked gloomy.

"It's okay! They were looking for a reason all along!" Samson assured.

Now, the time has come."

Samson called Margo over.

"Here's the plan! I'm going for the master... George?" Margo nodded. "Cover me!"

Fighters, take your positions! A dong went off ten seconds later.

Margo was in the lead with Samson right behind him! Leo and Suchi attacked at front while Samson and Margo fought their way through the middle. Tralina was taking down thieves with magic missile! Anna Trellis did intense damage with fire balls. Thor was back stabbing and slicing his way through the crowd. Thonolan swept away an arm and jabbed in the ribs.

Margo was getting close to the master who was slicing his way closer. He wanted this fight also!

"Brave man, taking on the likes of us!"

Slicing his way closer to the master, Margo slit a throat and slid his sword across a gut. Then he was close. Fighting to defend his dad, Margo stood firm.

When Samson reached Master George he revealed a platinum sword which he had smithed himself. He knew that it was sharpened to max.

Suddenly Master George called a seize fire.

"It's just you and me, Trellis!"

"You're on!" Margo sheathed his sword.

"To the death!"

"Agreed!" Samson relented.

The two men stood at guard. Master George took the first thrust. Samson blocked him easily! But then something strange happened... Samson's face began to glow! When it faded the fight continued. Every time George attacked Samson had a counter attack! Margo laughed... he knew exactly what had happened!

Samson fought George off and suddenly somersaulted over George's head! At the same time he slid his sword sideways and backstabbed him until the sword came out George's belly. George collapsed as Samson pulled his sword free! All fights were called off! The thieves were led away to be incarcerated. Their master was dead! This thieves guild would exist no more.

All the Avengers gathered together in one crowd!

"That was amazing! How'd you do that?" Samson's face began to glow again.

[Samson Trellis, do you accept this meld?]

"Indeed, I do!"

"Calibrating fingerprints! Let no one touch the Life Sword except he who wields it!] Samson's head glowed a little longer, then the light died.

"You mean..."

"I think I can explain!" Margo began. "When I sheathed my sword... dad revealed a sword of the same type! So, in seconds, Exceptor jumped from my sword to his! He is now the holder of the Life Sword! Exceptor's now with him!"

[It is advantageous that he used such a sword... it may have saved his life!]

"No, Exceptor, you saved his life! By just being there!"

"Wow, talk about power! I can see the whole continent in my head! Even Euta!"

Samson took a survey of injuries. Apparently, no one was seriously hurt. Leo approached.

"You made a risky decision going for the master, Samson!"

"But I knew I had protection! No stronger protection than a traveling spirit!"

Chapter 9

When the Avengers arrived back at the Estates Samson asked Leo to tell Sassy and Myna that Galala could return. He left immediately. They arrived half an hour later holding Anna's precious child.

"How'd it go?" Sassy asked.

"Master George is dead." Samson noted.

"Was anyone hurt?"

"Not a one!"

"How..."

[Congratulations on your pregnancy! It's a boy!] Exceptor informed.

"What?" Samson nodded agreement.

"You're carrying a son!" Sassy broke out in tears of joy.

"A son! A second son to take your place, Samson!"

"Only if Margo dies first!" Sassy couldn't help herself. She jumped Samson and kissed him deeply.

"Thank you! You're so sweet! I'll call him... Samson Junior!"

"Really?"

"Yes! Samson Trellis the second! It's perfect!" Samson drew his sword from its sheath.

"This is my Life Sword! Exceptor's essence lives here!"

"Wow, high quality! One of your own design, I see."

"How'd you know?"

"Look it the emblems on the handle. The falcon and the..."

"Oh, heck! I had forgotten about the crests."

"I'm gonna make love to you like never before!" Sassy assured. "A son!" Samson looked on enjoying Sassy's glee.

[Oh, Sassy. Don't worry... twins aren't in your future.]

"Really?"

[I don't see it any time soon.]

Sassy looked forward to delivering now that she knew it was a boy. Samson Jr., she would call him.

Into the eighth month Sassy grew impatient for her delivery. She was growing a stomach with each day that passed. On Galala's first birthday she was celebrating with the whole family. She suddenly noticed a difference in her son's movements. She was stunned into denial. But Junior just would not have it.

"Jr.'s coming."

"What?"

"Junior's coming! I'm getting contractions!" Sassy buckled up in pain. "It's starting."

Tralina Tralesta took her to a bed. She was undressed and prepared. Samson Jr. was born at 11:25 pm on Galala's birthday. Samson Jr was a beautiful Dwelven son! And healthy too! Sassy decided she would breast feed for as long as the milk held out!

Sasselia Wildlife Delivers Baby Boy

Sasselia Wildlife gave birth to a healthy Dwelven boy on Thursday. She calls him Samson Jr. Apparently he and Galala, Anna Trellis' daughter, have the same birthday! By only about 35 minutes. The Weekly Writing July 28, 1392

Leo Anasthasius To Marry

Leo and Samina will share their nuptials on August 3, 1392. Samina gave birth to a healthy baby girl on June 26th. Leo says he is not worried, they will try again! The Daily Scribe July 30, 1392

Samson called a meeting to order.

"As you know Leo is getting married soon and we will be there! I want to make the necessary arrangements immediately!"

Samson and his followers boarded the ship headed for Anasthasia without incident. Two days later they entered the palace and headed for the dance rooms and auditorium. Exceptor suddenly spoke up.

[Samson, we have a problem.]

"What kind?"

[Enemy kind!]

"What do you see?"

[Princess Amana... she'll be kidnapped!] Samson turned on his heels and headed for the personal quarters. When he arrived in the nursery he saw baby Amana asleep in her little bed.

[They're coming.] Samson hid behind the door.

[I've got something better!] Exceptor advised. Samson evaporated from all sight. The criminals were three thieves which Samson recognized from the coliseum. He drew his sword. Three minutes later all three were dousing the carpet with their own blood. Samson had beheaded each one of them.

[Good work!]

"It's just lucky you have precognition!" Samson replied.

[It's not over yet!]

"What you mean?"

[More thieves are waiting outside the palace!] Samson swore and headed back to the auditorium. At this point the bride and groom were on the pedestal. Samson ran up and stopped the service. He said a few words and then Leo and Samson both ran for recruits. The thieves were dealt with in a jiffy. Leo saw that each of them paid with their life.

"My daughter! They attacked my daughter! My flesh and blood! If it wasn't for you all hell would have broke loose!"

"It's Exceptor's precognition that saved her!"

"Well, I just thank God you're on my side! You've all become the most powerful entity on this continent!"

"Well, Exceptor did most of the work."

"Thank God for travelling spirits, that's all I have to say! Chaotics everywhere! They must have a new master!"

"Maybe one of the acolytes took over."

"This was too strategic to be... they planned this. They took a strategic attack against me."

"And now they're dead!"

"There's always more! They breed like hives! When some die others show up! There's never an end!"

"That's why you have us!"

"I hate chaotics!"

"That's exactly how I felt about Andrew!" Sassy notioned.

"You can't kill every chaotic on the continent."

"No... but it would help, wouldn't it?" Samson had no answer.

Amana Attacked By Thieves

The scheduled ceremony joining the royals of Anasthasia was suddenly interrupted by an attack on Amana, the royal princess. Report says that the Avengers along with faithful troops filed out and dealt with the attempted kidnappers. The Daily Scribe August 5, 1392

Chapter 10

Samson sat with Margo in the conference room alone.

"It feels different, doesn't it?"

"What's that?"

"Having two minds in you!"

"Yes, but you get used to it!'

"How is your wife?"

"Anna? Why you ask?"

"Have you been sharing sex?"

"Periodically!"

"You ever thought of trying again?"

"Another child?"

"Galala will be turning 2."

"Indeed, she will!"

"Do your day a favor! Give your daughter a baby brother." With that Samson left.

Margo found Anna playing with their growing daughter.

"Anna."

"Hi! Galala, daddy's here!"

"Um, Anna. I need you to listen."

"I will!"

"I think we should try again! Give Galala a sibling!" This brought Anna to full attention.

"You want..."

"I want... dad wants... mom wants."

"What do you want to try for?"

"You're interested?"

"Damn right, I am! I'll never give up babies!"

"You don't mind the birthing process?"

"Every woman forgets that as the child grows! It's our sacrifice for getting such beautiful beings."

"If we're lucky we could breed twins!" This shocked Anna.

"Twins?"

"Your mother did!"

"Everyone knows that twins skip a generation."

"That's an old wives' tail!"

"I guess. Well, when and where?"

"The north room."

"Upstairs?"

"Yeah! You can cast light so we can share!"

"Mom stays up there!"

"We can keep the door shut!"

"You're on! When?"

"Tonight! At dusk!"

"I'll meet you there!"

"You better watch Galala! She might become the next sorceress on you!"

"That would be my greatest pride! My daughter a sorceress!"

"Samson, daddy sent me a letter!"

Dear daughter:

Your mission is now activated. As head sorceress your mission is to enlist acolytes for the Territories! Search the towns... travel long distances... there are many Elves in Anasthasia. Many more travel Euta Vicinity according to Thor's latest news! The land beyond us is being populated quickly! Some go for a holiday, others go for adventure! Or just new things to see! Thor also says that Palapany birthed an offspring! Goldy named her Spirit!

Ludwig Wildlife
Elven Territories.

"What do you think?"

"I think we could make this a full time mission!"

"The Avengers searching for acolytes?"

"It's the ideal chance! Imagine the adventures we could experience!"

"There's one more thing! Daddy asked Leo to assist in any way he can."

"Then we're off to Anasthasia!"

"You really want to do this?"

"It's my heart's content!"

Margo went to the second floor. He found Anna in the main room. She was nude, he noticed. Her wraps were stored safely away. Moving to her he pinched a nipple. Then he took her in his arms and moved her behind closed doors.

Anna smiled her love.

"You didn't expect this, did you?"

"That you'd already be undressed?"

"Yes!"

"I love the way you giggle!" Margo stripped and sat next to her. As he sat his tool began to swell. Anna noticed. She bent close and kissed him. Then she grabbed his tool and he grew larger. She finally laid down and spread her legs. Margo looked her over and over again! He just couldn't wait any longer! He jumped her and bit a nipple as he growled. Anna remembered the first time he'd done that... at Galala's conception.

Margo entered gently and found her soft fur. He gasped once and she felt full and content being with him! He pulled and pushed half way. Then he pushed full! Anna gasped with excitement. Then the lust really began. For both of them! Margo pushed and pulled, totally out of control! Anna gasped her need. Margo pushed harder and suddenly sucked a nipple! Then he sucked harder. Hot flashes of lust spread through her. Then he let go and pushed his hardest! Anna cried out in happiness. She wet... then Margo released. Both rested, exhausted, never separating! Margo's tool flattened. But that's how Margo fell asleep... laying on top of her with his head on her breasts.

Chapter 11

Samson called a meeting to order the next day.

"We have a new mission. A primary mission which we will execute effective immediately."

"What is our primary mission?"

"I'm getting to that. Our new mission is to travel... find acolytes interested in learning the magical arts!"

"Male and female alike!" Sassy added.

"Male acolytes? It's never been done!" Anna reminded.

"Village Denvil has offered to accept males into a special temple! They will be trained as future Magicians and Sorcerers."

"That's really different!"

"Isn't it?"

"Why is this our primary mission?"

"Because as Head Sorceress it's Sassy's job to do just that!"

"Okay! How do we find these acolytes?"

"Grandpa Ludwig has said that a large population is moving into Galala Valley and beyond. Also, Anasthasia has quite a population of Elves."

"Understood. Where to first?"

"We will leave for Anasthasia first."

"By boat?"

"I don't think so! We go by land! No one knows who we might meet in our travels!"

"Well, you are Samson Leader, are you not?" Samson laughed out loud.

"We've been so busy preparing for tomorrow."

"I've arranged for Swina to keep Galala at her place!"

"I don't want to be away too long! Galala will miss me!"

"That's not a problem!" Samson suggested. "After Anasthasia I will separate the troops into assigned groups. Each group will travel their own way. I promise you, Anna, we won't be away too long! And as my daughter and law, I cannot even force you to come with us now!"

"Really?" Anna was all smiles.

"Stay if you want! Keep Swina company! She'd like another woman to talk to." Margo was watching the conversation.

"Anna, can I talk to you a moment?"

"Sure!" They walked out to privacy.

"I don't think you should come! We'll try again tonight and see. I think you shouldn't travel at this time."

"Because I might carry?"

"Yes! I know you have nine months but look what happened last time!"

"I do remember. Indeed, I do!"

"Let's keep our trying to ourselves... maybe some day you'll find me back with your tummy bulging!" Anna grinned.

"Tonight then!"

Margo and Anna did try that night again. The next morning Margo greeted Anna goodbye. Then he winked at her. She knew he meant to keep the secret to themselves.

The Avengers arrived at Colina Waterfront. They took a turn to the left and boarded the Colina Ferry which would take them to the opposite side of the river. There they would head south instead of west which would have taken them to the Dwarven Territories.

Two days passed. At last they arrived at the Anasthasia Ferry. It was currently in midwater coming their way. As they waited they noticed a boy and a girl, aged in their teens, Sassy guessed, coming their way.

"This could be a possibility for us!"

"Greetings, my friends! We are called The Avengers."

"I am Jacques and this is my sister, Mia."

"I notice you are Elven!"

"Indeed, we are."

"I want to ask you something! How'd you feel about apprenticing in the magical arts?" Both strangers stood stunned. Sassy continued.

"We are looking for potential acolytes... male and female alike! Village Denvil has offered to train males as Magicians and Sorcerers!"

"It's too good to be true!" Jacques declared. "I'm in."

"Me too!" Mia assured.

"We are still searching. We hope to collect many acolytes from Anasthasia."

"I can help you!" Jacques suggested. "The Elven population here keeps in close contact. We know almost all the Elven families in this region."

"Then we are lucky to have you in our troop. Ah, here's the ferry!" The Avengers boarded and soon were on Anasthasia Waterfront.

Chapter 12

"Where to first?" Samson asked of Jacques.

"I think we should head for Mrs. Strathman's. She's got a niece and nephew who live with her."

"Then you lead!" Jacques headed away from the Waterfront and led the Avengers through the residential area.

"Here we are!" Jacques knocked. A young Elven maiden appeared.

"Oh, Jacques! Please, you have to help! Auntie won't get out of bed! She's so cold!" Samson moved into the house and approached a bedroom. There was an old lady laying in a bed. Samson felt the body. Then he checked her eyes. He listened to her breathing. He discovered that she had died in her sleep.

"I'm sorry to say this, but your auntie is gone!"

"Then we're done for! She had our only income! How will we eat?"

"Come with us! We'll bring you to the Elven Territories. We are looking for potential acolytes. You can apprentice in the magical arts! Both of you!"

"I'm Chuck and my sister is Weena. Basically, we have no choice, do we?""Chuck, the scrolls!" Chuck moved away and brought back a box of parchment scrolls.

"80 years of Dwelven history. Auntie Tulie kept journals her entire life! And these scrolls contain everything concerning the history of the old Pledge of Allegiance. And Dwelven history!" Sassy gasped her surprise.

"Why did she write all this? It's neat and completely accurate. It even has the Dwelven birth dates and geneology."

"Our father was Dwelven." Chuck revealed.

"That explains why your aunt did this!"

"Auntie said she wanted this delivered to King Leo when she was gone!"

"Then we will do just that! We at least have to inform him."

"Let's go! I'm getting hungry!" Samson confessed.

The troop with extra Elves went directly to the palace. Samson found Suchi and handed over the box of scrolls.

"Everything you could ever want to know about Dwelven history!" Suchi checked out some of the scrolls.

"This is heavy stuff!" he revealed. "We'll consider this top secret! Give them to Sassy. She'll put them in a safe place!"Sassy jumped up and down!

"Really?"

"Take them. They have more to do with the Territories than they do in our principality."

"So, Chuck, Weena, you want to become acolytes?"

"At least we'll be fed! And it would be nice to have our own protection!"

"Indeed, it would!" Weena suggested. "Don't get me wrong, I wouldn't use your offer to our own benefit!"

"Jacques, who's next?"

"We're next! Time to find some food!" Chuck reminded.

The gang went to the Pig's Cleft Inn and ate their fill. Samson paid the bill. As they left they saw two Elves in trouble. They were two maidens who were fighting a thief. They were holding their own, but Samson knew they couldn't for long! He drew his dagger as his followers followed suit. When the thief saw he recruits he turned and ran.

"Let him go!" Samson insisted. "What happened?" he asked the two maidens.

"That thief tried to steal our travel bags! I pounded him with mine! It's heavy, but he was stronger. He never even budged."

"You're travelers?" Sassy asked.

"Ever since our mother died two years ago, we've been half way across the continent! I'm Sandra of Village Derlin. This is my sister Ana."

"Village Derlin, huh?"

"You look familiar. Do I know you?"

"I'm Sasselia of Village Wildlife!" Sandra stood stunned.

"You're the maiden of the Pledge! Sassy Wildfire, you name yourself!"

"Indeed I am." Sandra looked around at all of Samson's followers.

"Wait a minute... this is the Avengers troop! You're the Avengers!"

"Indeed, we are." Samson responded. Sandra stared at Samson.

"Your highness... forgive me!" Sandra bowed down to Samson.

"What are you doing?" Ana asked of Sandra.

"You're so nave! This is Samson Trellis! He's the king of the Dwarven Territories! He's royalty!"

"We're looking for potential acolytes! People willing to learn the magical arts!" Sassy informed.

"We used to be acolytes. We never finished our training! Our parents died and we gave up!"

"Would you like to continue your training now?" The two girls turned to each other.

"As long as we can learn in our home village!" Ana suggested.

"That can be arranged, I'm sure!"

"Six of them!" Samson reminded. "Maybe we should head for Colina."

"We promise," Ana insisted, we'll be a big help!"

Chapter 13

Myna Milner Trellis sat in the Boar's Head Inn with a flask of sweet wine between her fingers. She never even noticed when someone approached her table. Then she looked up.

"Can I help you?"

"You don't even recognize your own brother?"

"My..."Then Myna noticed his short stature. "Sonar?"

"The one and only!" Then he sat.

"What are you doing here?" Sonar ignored her and tested her flask.

"This is fine wine! Expensive, too!"

"Come on!"Sonar passed the flask back.

"Okay! I'm here to save your life!"

"What?"

"Why'd you leave Samson?" Myna flushed with his question. Then she spoke from her heart.

"Sassy needs him more!"

"You're saying you don't need him at all! That's what you're saying!" Myna broke out in tears. "You love him! He's your royal confidant! If it wasn't for him you'd still be childless! And, he's the love of your life!" Myna wiped away tears.

"Okay, I'm gonna ask you!" Myna suggested. "How do you feel when thirteen years of your fertility are wasted?"

"Where's this going?"

"Do you know how old Sassy is?"

"In her forties, maybe!"

"She's turned 44. She has six to ten years of fertility left! I'm only 39."

"So you gave up Samson for Sassy's sake."

"She's his first crush!"

"I'll make a deal with you!" Sonar suggested. "You go back to your husband."

"I have a career to think about!"

"Don't give that up! I'm not asking you to! Just move back to Avengers Estates and be there! Samson needs you! As much as you hurt to be with him!"

"You really think so?"

"He needs you... you need him... Sassy needs you too! Who's gonna be her midwife?" This brought a smile to Myna's face. "Just because Sassy is getting older doesn't mean you can cut off your own fertility! You're going to be an old lady with very few children! Don't give that up! And, don't ever give up your friends again! Now! I want you to trot yourself right back to Avengers Estates."

"Did you come all this way just for me?"

"No, actually, I'm on my way to the Elven Territories. Ludwig Wildlife is too old for headman anymore, so..."

"You're headman now!" Sonar nodded.

"You know me and Samantha Littleton have been married through the pledge... and she's the current sorceress... so that makes me headman."

"Well, I wish you luck with your new position."

"Remember, go home!" Myna nodded her promise.

When the Avengers and their new recruits got back to Avengers Estates they heard voices far off! Samson ran full trot to see who was here. When he got to the hallway he turned left. And there, in his sight, stood Anna and Myna Milner Trellis.

"Myna!" Myna cried tears of joy!

"I'm sorry, Samson. I never should have left!"

"I understand!"

"Do you?" Sassy showed up in surprise.

"Myna!" Myna trotted over and hugged her.

"I'm sorry! To all of you!"

"Why?" Sassy asked.

"I had a saint to clear my head! I talked with Sonar for a long time!"

"Sonar?"

"Her brother!" Samson informed.

"Sassy, I have to tell you! Sonar, my brother, he's taking over from your father as headman."

"Didn't I see a matrimony at the last Pledge Festival?"

"That was Sonar and Samantha Littleton!"

"Then with Sam as sorceress Sonar is the headman. There's no other choice!" Sassy clarified.

"You don't mind?"

"Not at all! Eventually they're gonna have to find a family of Wildlifes to take over for Sonar anyway! Village Wildlife isn't called Wildlife for nothing!"

"Who do you think will take over?" Samson asked.

"That's a very difficult question! Ludwig Wildlife only has daughters!"

"That's very true!" Samson agreed. "Let me introduce Jacques, Mia, Chuck, and Weena of Anasthasia. But these lucky two are Sandra and Ana of Village Derlin! And they want to go home!"

"Then we'll take them, of course!"

"Let's schedule a conference for tonight! By then I'll know where and who to send out for recruits!"

"I call this meeting to order! I have made my decisions as to who goes where! Atamar and Catamar, you go home to Euta Vicinity. Don't just search Calesta City. Go the whole way to Centaur Valley. Pick up any willing recruits you find!"

"Margo and Anna... you go back to Anasthasia!" Anna made a beckon. "Yes?"

"Samson, just so you know, I'm carrying!" This took everyone by surprise.

"You're pregnant?"

"Indeed, I am!" Samson smiled widely.

"I offer to watch her!" Myna suggested. "The least I can do for my soulmate."

"Okay, then it's Margo, Anna and Myna going to Anasthasia. Sassy and I will go as far as the Elven Territories. We will also search Galala Valley. Atamar, Catamar, you will join us as far as Galala Valley and then continue on! That's all! This meeting is adjourned."

Chapter 14

Samson and his troops went to Colina Waterfront with Myna, Margo and Anna to see them off. When they stepped onto the Colina Ferry Myna waved goodbye to Samson. He watched her sail as far as the eye could see and then turned headed for the Colina gates.

When the ferry reached land Myna stepped off and led the way. Two days passed as the family slept, traveled and ate. They lived on the old fashioned rations.

When they arrived at Anasthasia Ferry the boat was waiting nearby. They stepped on board. After they were seated two maidens stepped on board. They seemed friendly enough. Myan walked over.

"Hi, I'm Myna Trellis!"

"My name is China. My cousin here is Adana. We're from Village Denvil."

"Really? That's the village where magicians are going to be trained."

"Actually up until twenty years ago Village Denvil was known as Village Swena."

"Village Swena died out!"

"Died out?" Myna asked.

"When a village dies out it means all girls and no boys. The boys carry the name on. So when a village has no boys the village is considered dead!"

"So, if a village dies, what happens then?"

"Then the Elders choose a new headman. One who has male ancestors, preferably! Then the village is renamed. The old village dies and the new village begins!"

"We're here to search for potential acolytes. I see you two are skilled!"

"We are magic users! We learned quite a while ago."

Samson and his troops approached Village Wildlife. As they got closer they noticed Stewart Wildlife calling them to stop. He moved closer.

"Sassy, I'm afraid we have a serious problem!"

"What?"

"Your father... he's dying!"

"What?" Sassy galloped past him at top speed. Stewart observed the others.

"I'm sorry!" Samson led the rest of the riders at a trot. When they came to the village they noticed teary eyes everywhere! Samson felt sad himself! Dismounting he went to see Ludwig. As he entered Ludwig's bedroom he saw the old man lying silently. Sassy shook her father's shoulder. He awakened.

"Sassy..." Stewart showed up.

"What happened?" she asked.

"He suffered a heart attack!"

"Can he recover?"

"The medicine woman said there's nothing we can do!" Sassy cried tears of grief.

"Sassy..." Ludwig spoke again. "I'm old... and I'm ready... from now on... Village Wildlife is no more."

"What you mean, no more?"

"Village Wildlife is now Village Milner!"

"What?"

"Sonar Milner is now headman in my state! Village Wildlife is dead! And soon, so will I be!"

"Daddy?"

"Just think of it, Sassy! The first village to produce Dwelves! Village Milner will..." Ludwig shook and closed his eyes for the last time. Sassy didn't want to believe it. But there he was, her own dad... gone forever! She looked around noticing Sonar close by. Samantha waited in the living room. Steward led Sassy away as a sheet was placed over the body.

Sassy was still crying. Sonar stood stunned. Stewart moved to the office and obtained a scroll. He began to read from it.

"This is your father's will and testament. I, Ludwig Wildlife, as headman and leader of Village Wildlife, hereby declare that Village Wildlife itself is dead." This caught Sassy's attention.

"We have produced too many girls... there are no men to carry the family name! Village Wildlife is dead! And if you're reading this, so am I."

"My last will and testament is to see a village, a whole village, filled with Dwelven children. I instate Sonar Milner and Samantha Littleton as headman and sorceress of the newly named Village Milner! I declare that I have always loved the standards of the Pledge Of Allegiance. My biggest wish is to have an entire village of Dwelves! One safe haven for all Dwelves to live in together. That is my dream! May Village Milner live forever! This is my last will and testament."

October 5, 1392

Sassy stood crying and laughing at the same time. Her father had been so kind! An entire village of Dwelves! That had also been her own dream. A village of Dwelves to populate and alleviate stigma! Soon they'd have all sizes and shapes. All kinds of heights and all kinds of different features. Soon people would be looked at equally. Thanks to Ludwig Wildlife and the Dwarven-Elven Pledge of Allegiance! Soon her mission would be complete also!

Ludwig Wildlife Dies!

The high king and chief headman of the Elven Territories, Ludwig Wildlife, has passed away. It seems Village Wildlife itself is dead. In Ludwig's stead he has instated Sonar Milner as headman of the newly named Village Milner in honor of the Dwarven Queen Mother and ruling entity, Alanah Milner. It is said that Sonar's two brothers, Peter and Steven, will move to Village Milner and help in the population of Village Milner! The Daily Scribe October 12, 1392

Battle To The Death?

On June 3, 1394 the Elven Territories and fighters across the continent will join up to battle chaotics across the continent. The parties will be 2000 strong. They will meet in the newly released Galala Valley! It will indeed be a battle to the death! The Weekly Writing October 15, 1392

Chapter 15

Everyone was silent in the aftermath of Ludwig's death. But suddenly they heard someone come in.

"Who is it?"

"It's me!" Joseph Littleton, Sam Littleton's grown son, stood in the doorway. "I've come all the way from Littleton! Mom, I want to talk to you!"

"Sure!" Sam walked up. "What's up?"

"I know I'm twenty one, but..."

"Shoot!"

"I want to be an acolyte! I want to train as a magician!" Samantha looked at her son closely. There was determination in his eyes. Sassy strode over to talk but then she glanced away. Somehow, she couldn't look! Samson moved over to Sassy.

"What's wrong?" Sassy returned to crying.

"I'm sorry, Joseph, it's just that... you look so much like your father!"

"That's what everyone tells me!"

"I'm sorry I killed your father! And I'm sorry you were his!"

"So am I! I never knew him! In some ways I don't know myself!" Sassy cried harder.

"Auntie... I forgive you! It's not your fault I was born! It's not your fault that I don't have a dad!"

"Would you like a dad?"

"I'm too old now!"

"No, you're not! Sonar, meet my son, Joseph Littleton!" Sonar and Joseph shook hands.

"So you're my mother's new husband!"

"I am."

"I guess that makes you my dad. Or at least a step dad."

"Indeed, it would."

"Mom, he's okay! For a Dwarf!"

"Joseph!"

"Lay back! It's okay! I got a lot of Dwelven friends!"

"Really!"

"I'll bring them around some time! You know the women who were active at the first Pledge Festival? Well, their offspring are a little younger than me... but I've managed to get to know a lot of them... well, I'll go now. Grandpa's death is enough on our chests! See ya!" With that Joseph walked out.

"Sam, I'm sorry!" Sassy apologized.

"I understand. I really do!"

Myna stood at Anasthasia Waterfront.

"This has been a total waste of time." Anna suggested. "There is nobody to be found."

"That's not true! The waste of time, I mean."

"How's that?"

\ "We learned something today!"

"We learned about dead villages!"

"Well put, Margo."

"So, what do we do?"

"Only one thing to do... go to the Territories!"

"Catch up with the others?"

"There's no one here! We might as well!"

"So, how do we get there?"

"That's already taken care of!"

"What you mean?"

"Oh, they're coming!"

"Who?"

"The Elven Guard, of course!"

"Really?"

Stewart handed Samson a letter.

"This is what killed him!"

Dear Lawfuls and Neutrals of all the known continent: This is a notice to join us in a battle to the death! You will meet us in Galala Valley on June 3, 1394. You will carry 2000 heads. As we will hold 2000 heads. If you do not comply on the chosen date we will ransack your homes, kill your children and steal your wives! Your villages will be burned down and you will still be killed! This is the final warning! Meet us or suffer the consequences!

Chaotic Thieves Guild

"Wow! They're cruel, aren't they?"

[It seems the war is on again!]

"Can we find 2000 heads?" Sassy asked.

"I can call on my Dwarven Soldiers!"

"I can call on my magic users!"

"How many is that?"

"1500 heads."

"Still 500 short!" Sassy calculated.

"What about Leo's fighters?"

"200 heads."

"What's the date today?"

"January 10, 1393."

"Sassy, we have to arrange the funeral!"

"Of course!"

Chapter 16

Myna and her followers arrived in Village Wildlife, but not a soul was seen. They trotted on horseback through the village and beyond. The saw no one! Then Myna heard a soft voice in the distance! It lead her to the Elven Temple. She dismounted and led the way. The entire town seemed present. She walked the floor and sat next to Samson.

"What's going on?" she asked.

"Ludwig's funeral!"

"No!" Samson nodded.

Samantha cast a fireball onto Ludwig's casket. The lid buckled and fell in. The flames would continue to burn until nothing would remain but dust! Sassy started the eulogy.

"My father lived his life well! He helped reinstall the Pledge of Allegiance. He accepted and nurtured my sister, Samantha! He loved his family! And we loved him!"

"His greatest achievement ever is now underway! Village Wildlife is indeed dead! This new Village Milner will be a fresh start! A new beginning! It will be a safe place for Elves, Dwarves and Dwelves alike! God Bless Ludwig Wildlife!"

After the funeral Samson and the rest of the family returned to Sassy's. Samson showed Myna the scroll from the chaotics.

"Another war?"

"Remember, Palapany said the war would reach Galala Valley! This may be the war she was talking about!"

"A war that can't be stopped! But 2000 heads!"

"We have 1700 available."

"Where do we get the other 300?"

"We train them!" Sassy suggested. "We have more acolytes training right now! With all the villages together... probably 400 acolytes are training. 300 or more of them are graduating in about 12 months! We could get the required numbers! Because we have a whole year!"

Samson called a meeting at Village Milner.

"I call this meeting to order! I have decided we will continue our primary objective! Catamar and Atamar will continue their trip. They will search Galala Valley as well as Euta! Myna, you have a report?"

"We met two skilled maidens but no acolytes."

"Then we stand down! We don't need to return to Anasthasia!" Margo spoke up.

"Dad, the maidens were from Denvil! We could use them!"

"That's something to consider! In that case, your team will search them up! Meanwhile, the rest of us will return to Avengers Estates."

Three days later the teams split up, Myna and her team headed for Anasthasia, and Samson's group would return home!

Myna stepped onto Anasthasia Ferry. They were alone. When the boat reached Anasthasia Waterfront they stepped off and scanned the area. The two maidens were nowhere to be seen. Myna turned to a sailor and asked.

"Have you seen two maidens? China and Adana Denvil?"

"Denvil? Two ladies with that name took a room at the local Inn. But they're checking out soon! You might of missed them!"Myna called the others and they headed for Pig's Cleft Inn.

When they arrived they approached the bar.

"I'm looking for two maidens named Denvil! I need them!"

"They said they'd check out soon! You caught up just on time! Room number 4." Myna took the stairs. She knocked on door number 4.

China answered the door.

"Hi! We meet again!"

"I've been looking for you! I need your assistance... actually, we need your assistance!"

"What's up?"

"Chaotic thieves have challenged us to a duel! In Galala Valley! We need your skills!"

"We're actually packing up headed home! To Village Denvil!"

"Will you help?" China turned to Adana. She nodded yes.

"I guess we're in!"

"When are you leaving?"

"Immediately!"

"Can we travel with you? We're going home to Colina."

"Then you're welcome!" The two maidens checked out and Myna arranged a ride with the Elven Guard.

Chapter 17

"Two thousand heads? This is a big responsibility!"

"I'm still in!" Adana confessed.

"You're sure?"

"China, we're doing this for the Territories! For everyone!"

"You're sure!"

"I'm in."

Myna had been explaining the situation during their trip to Colina.

"Then I am too!"

"Done!"

The Elven Guard soon set down at Colina Waterfront. Everyone got off. The maidens told Myna that they must continue their travels but would show at Village Milner on time.

Myna walked into Avengers Estates followed by Anna and Margo. Anna had been sick to the stomach but Myna assured it was normal.

Samson made a meeting for the same night.

"I call this meeting to order." Myna spoke.

"China and Adana both agreed to meet at the chosen time!"

"Then that's settled! I received a letter from Catamar. He's found three maidens touring Galala Valley! They're all acolytes at Village Sneela on the edge of the Elven Forest. But they're all first year acolytes! They start their second year in February. We might be able to use them after they graduate! Catamar also says he is continuing on to Calesta City. He says that this will be his last possible contact!"

"Did he state any first names?"

"Andreena, Rachel and Seeno. All from Village Sneela."

"Samson."

"Yes?"

"What do you think of visiting Village Denvil? The magicians are being taught there. We should pay them a visit."

"I'll schedule that for later in the year! Right now Galala and Junior need us. We'll talk about that later!

January flew past and February afterward. In March Anna carried with a bulging tummy. Just a few months left.

Samson spent most of his time at the Boar's Head Inn talking to Peter. Peter was keeping a close watch on Dwarven politics. Later that month Samson wrote a letter to Alanah.

Dear Alanah, I greet you and considering your situation as royalty I am sending my son, Margo Trellis, to rule in your place! With him will be Anna, his wife, and new Queen of the Dwarven Territories. I hereby release my crown to my son, Margo Trellis! I look forward to seeing Margo's skill as King and leader of the Dwarven Territories! Yours, Samson Trellis March 18, 1393

Samson called a meeting to order.

"I haven't talked to any of you about what I'm about to do! But it is necessary!"

"What's up?" Sassy asked.

"Starting now I release my crown as King of the Dwarven Territories! From now on Margo Trellis will be King in my place! And Anna Littleton Trellis will stand as Queen!"

There was complete silence.

"Margo? Will you accept this honor?" Margo stood stunned.

"These last years I've been an Avenger! I've been one of you! But dad's right... Alanah can't rule indefinitely. I have no choice but to leave you and take my place as king. But why now?"

"Like you said, Alanah can't rule indefinitely. Also, considering the need for Avengers I cannot return! I will lead the Avengers... you will lead the Dwarven Territories! Will you accept?" Margo turned to Anna. She had a wide smile on her face.

"Take it! I want this too!" she assured.

"The crowning will be held soon! Meanwhile, I'm ordering that all of us head for home! On the morn!"

Some six days later Samson approached Trellis Settlement.

"Margo, you and Anna will live here!"

"Really?"

"And Alanah will be your Nanny."

"Wow!"

"Hi, you all!" Alanah was already there. She was aged at 68. Her face was wrinkled a little bit but the brown eyes still twinkled. Myna gave her mother a big hug.

"Did the boys leave?"

"They moved to Village Milner. Yes, they left last month!"

"Anna, Margo, welcome home!" Samson offered. "Margo, you'd best work as the local smithy! Don't want to lose our family trade, you know!"

"I promise! I'll take care of Trellis Settlement! And I'll continue platinum smithing!"

"Well, we have the crowning to think about!"

Margo Trellis New King

Samson Trellis has released his control of the Dwarven Territories to his son, Margo Trellis! The crowning is scheduled for tomorrow. The Daily Scribe March 28, 1393

Margo sat in the King's chair. Anna was seated beside him. The entire Territories seemed in attendance for this special moment. Samson approached with the King's crown. Sassy held the Queen's crown. Samson placed his crown on Margo's head! Then Sassy put the Queen's crown on Anna's. Samson moved in front of them.

"Margo and Anna Trellis, will you accept this position as the new royals of the Dwarven Territories?"

"Indeed we do!" they both assured. Samson turned.

"Ladies and gentlemen," he yelled, I introduce King Margo and Queen Anna. They were married under the standards of the Pledge of Allegiance. I await your acceptance of said royals. Do you accept them as your own?"

There was thunderous applause! Then two men appeared. Leo Anasthasius and Suchi Evil Killer walked up the stage. Leo spoke!

"Ladies and gentlemen, as King of the Principality I sanctify and honor this crowning. May the Dwarven Territories live forever!"

Chapter 18

Margo's first role as King was to prepare soldiers for the coming war. He also decided to hire one of Samson's own generals to start training new recruits! He chose Satira Anders! Satira was aged at 91 but Margo still insisted. Satira was still capable and healthy. Satira finally relented.

Samson and his troops had returned to Colina. Catamar and Atamar arrived back in September of 1393.

Meanwhile, Anna birthed male Dwelven twins on June 28, 1393. However, the twins were not identical.

I call this meeting to order. On the agenda... Catamar, I await your report."

"We found six Elven maidens and two male acolytes. We did as you asked... we went all the way to Centaur Valley... on the way we picked up the acolytes in several different places! We found one of the males living at my old cave! He offered to be there when he was needed. He's currently at Village Denvil. As all of them are currently in training."

"That's assuring."

On May 31, 1394 Samson and his troops arrived at Village Milner. The town was crowded. He set his position and the 2000 odd party headed toward Galala Valley.

On June 2, 1394 Samson arrived at The Gate. Standing in front of them was a beige colored horse. Suddenly air shifted and there stood a strange beige Centaur! Samson strode closer.

"I greet you, honorable Centaur!"

"I greet you, Samson Trellis! I am known as Spirit, I am the direct offspring of Palapany."

"Thor has mentioned you!"

"I am currently the keeper of Galala Valley. And yes, indeed, this is the war my mother mentioned. It cannot be denied."

"Then we will continue on our way! Stay safe!" Samson travelled til dark and set up permanent camp.

The next morning Samson saw horses carrying leather armored thieves in the far distance. Then Samson whistled. Spirit appeared out of nowhere! Samson climbed on Spirit's back calling Sassy to join them. Then Spirit took off!

Spirit ran all the way to the Galala Shrine. Then Samson and Sassy entered the shrine. But then Samson began to glow. Exceptor left the life sword and became a light on the ceiling.

[I brought you both here for safety. What I'm about to do is horrendous but necessary! After this I will be forced to return to the Underworld! Watch!]

Both Sassy's and Samson's heads began to glow. They suddenly saw a vision of the troops. The thieves were setting stance. The lawfuls followed suit. On a count both parties ran toward each other. They moved close, and just before they clashed something strange happened. The sky was grey. A strike of lightning started a grass fire. But it headed in the direction of the chaotics! The lawfuls were safe! The flames increased with the dry, long prairie grass! The chaotics began to panic! Some ran! Others caught fire before they could do anything! The ones that ran were ahead. But suddenly another strike lit the grass ahead of the runners. Now there were two grass fires! The runners stood at a loss.

The ground began to shake and a huge quake opened! Both fires were coming toward them! They decided they didn't have a prayer! Some killed themselves! Others jumped into the quake! Still others caught fire and burned to death!

In the end the chaotics were defeated. Sassy and Samson returned to their own states!

[Forgive me! But I will not be forgiven! There are consequences for these actions. I cannot stay! I must return to the Underworld!]

"Did this really happen?"

[Indeed, it did!]

"How'd we see it?"

[I've melded with both of you!]

"Not with me!" Sassy suggested.

[You've forgotten the Underworld! I melded with you there!]

The light began to glow and then a bodily form appeared.

[This is me. My name is Arthur Huntingham, I am a human magic user, and I come from Anasthasia.]

Exceptor looked to stand six feet one inch and had blue eyes, a stand out nose, black hair and buckled muscles. He wore a black robe.

[Forgive me! And remember me! Your minds will now be free!] Samson stood clearing his mind. The essence of Exceptor was gone! Samson left the shrine and saddled on Spirit once more. When they finally arrived on the front lines he could smell burned flesh in the air! The dead were dealt with and everyone gloried in their safety! No one knew what had started such weather... or how the miracle had happened... but it was considered simply an act of God!

Samson called to pack up and leave! The continent was safe! Soon Samson arrived back at Village Milner! Samantha Milner had some good news! She was expecting her second child. Hopefully it would be a boy!

Samson went back to Avengers Estates to find Leo Anasthasius present. He turned to Samson.

"Some would have called that catastrophe a work of God... but I know better! It was Exceptor, wasn't it?"

"Yes, it was! But he's gone now! Back to the Underworld!"

"Yes! We'll be just fine without him!"

"Will we?"

"Look at it this way... Margo's king! Village Milner is gonna grow and life will continue. For chaotics too! I assure you, they will show again!"

"I'm counting on it! Because, that's what we're here for! Us Avengers!"

Epilogue

With Margo Trellis as king and Samson as head of The Avengers, the thirteenth century would go out in a storm! Village Milner would grow at its own rate! Sassy Wildfire and Samson Trellis would head The Avengers together and maybe, some time very soon, chaotics would renew! Because, indeed, that's what The Avengers are for.

The Dwarven-Elven Pledge Of Allegiance

In my writings of my fantasy series the Dwarven-Elven Pledge Of Allegiance is rampant throughout. The Elves and Dwarves in this series have the same ancestry! Some time in the past these two nations were separated from each other because of height! The stigma rose against interbreeding! At one time the very thought of an Elven wife brought Samson to shame. The Elves and Dwarves stuck to their own kind. The Pledge brought prosperity both for alleviating stigma and the financial status of both peoples.

The mutants of Colina were not mutants at all! That is another kind of stigma! The mutants turned out that way simply by interbreeding.

The fact that Samson could fall for an Elven maiden itself is proof that size, race or creed should not and cannot be considered a factor!

I believe that if you can get to know what a person is thinking at any given moment, then you know them well.

A relationship takes five years of celibacy to earn their love, trust and friendship.

The Pledge was also a way for nations to mix, meet new people and socialize!

The Dwarven-Elven Pledge of Allegiance is just one example of people accepting people for what and who they are!

Biography of the Author:
Marcel Rene Chenard, A.K.A. Sassy Wildfire, was born on June 14, 1967 at Spirit River, Alberta, Canada a French Albertan. He was born with low blood sugars and suffered from epilepsy until his fourth year. In school his best subject was always English. In 1986 Marcel contracted Schizophrenia. In 1991 he lived in Edmonton and then began writing books with the help of Austin Mardon of Golden Meteorite Press. Marcel's first publication was called *Lady Juanita Rodriguez.* Soon afterward Marcel began writing Fantasy novels based on the Dark Ages or thirteenth century. Marcel wrote under the pseoudonym Sassy Wildfire for the Fantasies as this name is rampant throughout the series. Marcel's fifth and last fantasy novel, *In The Aftermath*, is currently being processed. He uses the name Sassy Wildfire for many of the books he writes; not just the fantasies. Marcel Rene Chenard A.K.A. Sassy Wildfire currently lives in Grimshaw, Alberta where he is looking to a new dirrection for his written works.

www.ingramcontent.com/pod-product-compliance
Lightning Source LLC
Chambersburg PA
CBHW021938220326
41599CB00010BA/523